CELEBRATING URBAN COMMUNITY LIFE

Fairs, Festivals, Parades, and Community Practice

Communal celebrations bring out the best in us, offering a place for people to come together and take a break from the routines of daily life. They are a vital aspect of city life and are increasingly popular as an urban development strategy.

Celebrating Urban Comm  ide to understanding and enhancing ce of community capital. Drawing on c 'rancisco, and Toronto, Melvin Delgado ch fairs, festivals, and parades can en a framework for social scientists, urban planners, and social workers to analyze and foster celebrations that benefit urban populations, the book is a valuable resource for those with an interest in this growing area of academic and practical interest.

MELVIN DELGADO is a professor in the School of Social Work at Boston University.

MELVIN DELGADO

Celebrating Urban Community Life

Fairs, Festivals, Parades, and Community Practice

UNIVERSITY OF TORONTO PRESS
Toronto Buffalo London

© University of Toronto Press 2016
Toronto Buffalo London
www.utppublishing.com
Printed in the U.S.A.

ISBN 978-1-4426-4995-8 (cloth)   ISBN 978-1-4426-2748-2 (paper)

♾ Printed on acid-free, 100% post-consumer recycled paper with vegetable-based inks.

Library and Archives Canada Cataloguing in Publication

Delgado, Melvin, author
Celebrating urban community life : fairs, festivals, parades, and community practice / Melvin Delgado.

Includes bibliographical references and index.
ISBN 978-1-4426-4995-8 (bound). – ISBN 978-1-4426-2748-2 (paperback)

1. City and town life – Case studies. 2. Cities and towns – Social aspects – Case studies. 3. Festivals – Case studies. 4. Fairs – Case studies. 5. Parades – Case studies. I. Title.

GT3930.D44 2016   394.2609173'2   C2015-908257-9

University of Toronto Press acknowledges the financial assistance to its publishing program of the Canada Council for the Arts and the Ontario Arts Council, an agency of the Government of Ontario.

This book is dedicated to Denise, Laura, and Barbara

Contents

Acknowledgments ix

Section 1: Setting the Context

1 An Overview of Urban Community Practice   5
2 An Overview of Community Celebratory Events   25
3 Evaluating Community Celebratory Events   63
4 Capacity Enhancement of Community Assets   84

Section 2: Community Celebratory Events

5 Fairs   101
6 Festivals   114
7 Parades   130

Section 3: Case Examples of Community Celebratory Events

8 Case Example of a Fair: NYC Chinatown's Health Fair   143
9 Case Example of a Festival: San Francisco Chinatown's Lunar New Year Festival   154
10 Case Examples of Parades: Toronto's Portuguese, Hispanic/Latino, and Pride Parades   164

Section 4: Reflections on Community Celebratory Events

11 Knowledge and Competencies for Community Practice and
   Celebratory Events   179
Concluding Thoughts   188

References   195

Index   277

Acknowledgments

A book is never the work of one individual, and this book is no exception. I particularly want to signal out the contribution of Christine Lee, Boston University School of Social Work research assistant, and her work on developing the case study of the Chinese Lunar Year Festival. Her diligence and insights were outstanding.

CELEBRATING URBAN COMMUNITY LIFE

SECTION 1

Setting the Context

1 An Overview of Urban Community Practice

Celebrations of any kind bring out the best in us and those around us. Many of us use celebrations as distinct markers in our lives, further increasing their importance. Whether it is a party, festival, parade, fair, or some other social event, celebrations provide a place and space for us to come together to connect, share, eat, laugh, possibly sing and dance, and have a break from the routines of daily life. The popularity of such events is such that a book in the popular series of "dummies" was devoted to them (*Meeting & Event Planning for Dummies*, Friedman, 2003), and almost 15 years ago! Although this book focuses on business and trade shows, the contents share much in common with the "business" of celebratory events.

The following quotes captured by the *New York Times* reporters Yee and Turkewitz (2014, A12) at New York City's Puerto Rican Parade in 2014 illustrate a variety of reasons for attending this community event, from celebrating history to honoring the community's achievements. One 26-year-old who was there with his two-year-old son said, "I used to come here every year with my grandfather, and I'm just carrying on the tradition." One 66-year-old had attended 27 of these parades. One of the parade's grand marshals was the recently elected City Council speaker, who was born in Puerto Rico; because of her new role, the parade was extra meaningful for her: "The community is celebrating my accomplishing being the speaker of the City Council — it's a great recognition for the community overall. And … we've taken this parade back to its roots." These statements capture a range of motivations for attending and give context to understanding celebratory events (Jayaswal, 2010).

Attendance at celebratory events is rarely an individual experience and often involves going as a group of family, friends, or some combination of these. The typical social group attending a festival is up to six persons (Oberhagemann, Könnecke, & Schneider, 2014). Attending as a group creates a significantly different experience and memory compared with attending as an individual. When groups come into contact with each other, we have the makings of a crowd.

Fine (2014) proposes a "meso-sociological" approach toward groups and civil society, which can help increase our understanding of the function of urban celebratory events, and identifies six concepts within that approach: the culture shared by a group helps them obtain commitments from members; social capital and social relations can build capacity and efficacy; public places can help create shared spaces; public performers can bring residents together to take action; shared histories can contribute to a unified vision of a future; and social control can function to regulate behavior and stabilize interactions. Fine's themes address many of the elements associated with celebratory events and why they can transform communities.

A literature review of community celebrations highlights how such events have provided an analytical portal or window for the study of public space, ritual, power, opposition, and community (Foxall, 2014). These events are complex social phenomena, serving many different functions, breaking up the mundane, interjecting gaiety, and relieving us from the trials associated with daily life. New Year's celebrations, for example, have taken on extraordinary meaning throughout the world (Montgomery & Christie, 2011), as many of us either participate in or watch the festivities on our televisions, embracing the symbolism of the event.

Loukaitou-Sideris and Ehrenfeucht (2009) discuss how celebratory events disrupt the flow of everyday life in an affirming and participatory manner. Ottawa Chinatown's Lunar New Year Lion Dance typifies how sections of a city are transformed when a community puts on an event (Ottawa Chinatown, 2014). Celebratory events transform public places in ways that can seem magical, creating new experiences to possibly replace negative ones. Further, the "magical" outcomes of events create collective memories for a community and connections that last long after the event is over, becoming part of a community's lore. These events have tended to be relegated to our personal lives, yet this does not have to be the case (Koch & Latham, 2012). Fostering such events can be part of community social work practice, although, with few exceptions, community celebratory events have not been conceptualized in this manner within the field (Reverte & Izard, 2011).

However, the popularity of planning and executing celebrations is not restricted to event planners or community organizers; "ordinary" residents are probably responsible for most community celebratory events. Although academic institutions have played a role in helping communities plan celebratory events (for example, the North Dakota State University Cooperative Extension Service published a booklet entitled *Planning Community-Wide Special Events* [J.P. Smith, 1986]), the academy is playing catch-up regarding urban community practice and celebratory events. Qadeer & Agrawal (2011), for instance, point out that theory lags behind practice in the realm of planning "multicultural" events (activities that affirm and build on the presence of minority groups) in Canadian and American cities.

A search for urban practice interventions, such as geographic-specific social planning and program development, is ever constant as the world's cities are growing in number and size, and it is projected that they will continue to do so for the next 40 years (Colding & Barthel, 2013; Huang & Zhao, 2013; United Nations, 2013). Cities are increasingly diverse and bring challenges for how visible minority communities can develop identities that are self-affirming while still being connected to the broader society (Berg & Sigona, 2013; Chan, 2013; Hou, 2013; Loukaitou-Sideris, 2012; Poirier, 2011). As such, urban practice that affirms and engages marginal groups is highly desirable and open to innovative processes. However, Zapata-Barrero (2014), on the basis of experiences in Barcelona, Spain, emphasizes the challenges with achieving participatory goals when bureaucracies are involved and noble intentions are subverted.

Segregated communities lead to social, economic, and political tensions and problems, necessitating innovative interventions to address these circumstances and the resulting damage to the communities' identities (Inglis, 2011). These interventions must be grounded within the cultural values, history, and traditions of a group. Over the past several decades various forms of community practices have been addressing these challenges (Beard & Sarmiento, 2010; Gapas, 2013; Heywood, 2011).

The search for innovative interventions has generally focused on finding solutions to "urban ills" and has disregarded community assets, including cultural heritage. This negative viewpoint has come to be known as the *deficit paradigm*. Butticci (2013) identifies terms often used to describe Lagos, Nigeria, which highlight this very point: "urban apocalypse," "anarchic urban catastrophe," and "urban dysfunction."

The deficit paradigm or perspective buys into a highly destructive bias that society has toward cities. Kazin's essay "Fear of the City, 1793–1983"

(2008) traces the history of how urban America has been viewed and feared over almost two centuries and the forces that have shaped these views. Even today, newspapers rarely extoll the virtues of an "urban oasis" or "urban sanctuary." Instead, it is more common to hear about urban "crime," "plague," "distress," "violence," "disease," "poverty," and "drug problems." With the exception of "rural poverty," rural is usually not associated with problems, and suburban is simply missing in this discourse, although issues of "suburban poverty" are starting to find currency. Urban bias is prevalent and complicates making a shift from focusing on deficits to focusing on assets in a community. Recognizing the prevalence of this bias is a critical initial step in recognizing the value of urban celebratory events.

Any review of media coverage of urban communities reveals a very depressed state of existence, with seemingly little for residents to be proud of or look forward to (Marquez, 2012). Governmental statistics further reinforce this negative image by focusing on deficits (Jackson et al., 2003). Nevertheless, there are glimmers of hope, and even a congratulatory stance, to be seen in various community celebratory events, highlighting a potential source for innovative community practice that is not only affirming but also improves a community's overall well-being by enhancing its social capital, or assets (Long, 2013; Lopez-Bonilla et al., 2010). Capturing these assets means we must first acknowledge that they exist and then specifically search for them. Asking the right questions is critical; for example, "what are the glimmers of hope and dreams?" versus "are there any?" Starting with a positive assumption shapes how we proceed and the language we use. Furthermore, the narrow and distorted view of urban communities inherent in the deficit paradigm raises ethical questions pertaining to assessment and practice. No community consists exclusively of deficits, but the deficit bias leads to policies and programs that systematically undermine community assets, social identity, and the ability to thrive (Daniels, 2010; Magis, 2010).

There are challenges facing low-income/low-wealth urban residents, and one is never at a loss to find a plethora of statistics attesting to that fact. Nevertheless, there are assets, too, that often get overlooked by scholars and the popular media which systematically paint a picture that is hopeless for these communities. And unfortunately, there is no corresponding plethora of stories and statistics on assets (Delgado & Humm-Delgado, 2013).

Urban communities with multiple jeopardies are especially hard-hit. Focusing only on these deficits can mean that valuable insights and opportunities are missed for identifying and mobilizing a community's assets for community-based projects and benefits, and for affirming the community's positive attributes. Instead, a community's story becomes distorted to the detriment of its residents and society as a whole. Community social work practice can help communities develop a different, more positive, narrative. Community celebratory events can be conceptualized as community storytelling enhanced by practitioners (Little & Froggett, 2010).

Mair and Whitford (2013, 16) address the importance of celebratory events, and in the process, help set the stage for this book: "Throughout the centuries, a multitude of events have been staged as fairs, festivals, sporting and cultural events, exhibitions and business events (among others). Events have played, and continue to play a significant role in society ... Events provide a means of marking important personal and public occasions and celebrating important milestones in our lives. Celebratory events are like an elixir ... with the ability to facilitate social capital as they bind communities [and] facilitate constant rejuvenation of the community experience." Mair and Whitford's assessment sheds light on how celebratory events shape individual and community collective memories; however, these collective memories can also be contested memories (Robbins & Robbins, 2014).

Urban Community Life as a Backdrop and Context for Celebratory Events

It is impossible to discuss community celebratory events without discussing the concept of *community* since it is the contextual setting (Cutchin et al., 2011; Edwards, 2011; Smith, Bellaby, & Lindsay, 2010; Zukin, 2010). Defining community may seem like a simple task, but nothing could be further from the truth (Chaskin, 2013; J.W. Murphy, 2014).

Community is so much more than geography, and Gieseking and Mangold (2014a, 73) make this point: "Why do we feel that we belong in some places and not in others? Place and identity are inextricably bound to one another. The two are co-produced as people come to identify with where they live, shape it, however modestly, and are in turn shaped by their environments, creating distinctive *environmental autobiographies*, the narratives we hold from the memories of those spaces and places that shaped us" (original emphasis).

There are countless ways of conceptualizing urban community and public spaces, with each profoundly influencing how best to intervene in the life of a community (Chiodelli & Moroni, 2014; Milbourne, 2012). Vukov (2012, 129) notes that urban ethnology offers the promise of helping academics and practitioners understand spaces, enhancing our understanding of the meaning of celebratory events: "Issues related to city space, its main locations, and the ways in which they are integrated in the lives of urban communities form an important part of research in urban ethnology. Studying the city from the perspective of different elements of its topography (streets, squares, neighborhoods, markets, etc.) has been stirring a vivid interest among ethnologists, sociologists, and anthropologists."

Urban topographies are places and spaces where cultural differences are negotiated, connecting residents from disparate backgrounds and providing the context in which celebrations occur (Koefoed et al., 2012). Public spaces permit public participation in democratic societies, and these social exchanges can achieve a high level of importance (Marcuse, 2014).

Some scholars argue that the concept of community has outlived its usefulness and that we are best conceptualized as living in a postcommunity society. The concept of cyber-community has emerged to substitute for community in the conventional sense. Nevertheless, a search of the literature finds that *community* is still alive and well; there are numerous scholars who would argue that *community* is thriving with its demise nowhere in sight, although it remains controversial how best to define and measure it (Jeffres, 2010; Kuecker, Mulligan, & Nadarajah, 2011; West, 2012).

Community as a focus and vehicle of social change has a rich and long history (Ginwright, Noguera, & Cammarota, 2013; Shragge, 2013). Community is ubiquitous and has been a rallying cry from proponents and opponents alike in crafting social interventions (Sites, Chaskin, & Parks, 2007). There is a close relationship between how a community is defined and how capacity enhancement is operationalized using the resources (capital/assets) that can be identified and mobilized in supporting communities and celebratory events (Betancur, 2011; Norton et al., 2002). Bringing together community, capacity enhancement, and celebrations complicates our understanding of celebratory events but it is essential that we do so. Nam (2012) discusses how the youth in Chicago's Puerto Rican community can be engaged in working in community gardens and at celebratory events, providing an opportunity (both space and place) for learning about their history and enhancing their competencies in the process. Flachs (2010) focuses on community gardens and celebratory events in Cleveland, drawing attention to how public spaces play a role in fostering the social fabric and shaping interactions within a community (bonding social capital).

The relationship between community and events can be viewed from a multifaceted and multidisciplinary perspective (Chapple & Jackson, 2010). Positive community cohesion (quantity and quality of interactions between residents), or social capital, has been associated with celebratory events (Litman, 2012). Beider (2011) calls for a new framework for understanding working-class communities; while his report focuses on communities that are white and English speaking, the same call can be made in other Western communities. Procter (2006) refers to community-coalescent events, or events and activities that bring residents together in an affirming and participatory manner, creating spaces for groups to gather and form identities.

Achieving high attitudinal consensus represents another view, and it is an important initial step in creating effective collaborations among neighborhood associations (Hyde & Meyer, 2010). Small-scale community events can increase social capital (bonding and bridging) and personal efficacy, and this can translate into other community benefits (Molitor et al., 2011).

Social connections emerging from attending celebratory events are considered major positive outcomes. These new relationships (which can be intergenerational, intercultural, and, in the case of newcomers, between groups with different levels of acculturation) bring depth to achieving increased social capital as a goal. The creation of an opportunity to form new relationships, and the breaking down of stereotypes, can transform into hope for a unified community with shared goals. Celebratory events are rarely so specific regarding goals for new social relationships resulting from attending. Nevertheless, this can occur.

Events need to be grounded in space and place (Gieseking & Mangold, 2014b, 3–5): "Space and place are co-produced through many dimensions: race and class, urban and suburban, gender and sexuality, public and private, bodies and buildings ... the relationships between people and place have many facets – psychological, social, physical, and cultural – and include structural and institutional forces that may originate in distant places or times." Celebratory events are attractive vehicles for helping communities achieve and maintain high levels of new relationships (bonding and bridging social capital), including increased levels of trust, personal efficacy, and identity formation and reinforcement, which contributes to building a community consensus (Onyx & Leonard, 2010).

These events have the potential to increase tolerance of differences (Bannister & Kearns, 2013). Community events that actively reach out to community outsiders can engender exposure to segments that they normally would not interact with, helping dispel stereotypes that outsiders might have of the community, and increasing tolerance of strangers (Wu & Pearce, 2013a). The need for increased tolerance is not narrowly restricted to attitudes toward newcomers or different ethnic and racial groups, and can include LGBTQ (lesbian, gay, bisexual, transgender, and queer) groups (Kondakov, 2013), as well as attitudes within these groups. For example, Whitesel and Shuman (2013) focus their attention on Girth and Mirth (G&M), a social group of "big gay men," and their exclusion from meaningful participation in public activities sponsored by the LGBTQ community.

There are many forms of community events that are not typically classified as community celebrations, and there is always room for innovation when addressing community events. Cumming (2012), for example, examines a community project where a play was developed based on historical letters to the editor in local newspapers. Fairs, festivals, and parades bring elements that make celebratory particularly attractive for urban-centered community practice.

Various academics have looked at community celebratory events from different angles. Social scientists, particularly anthropologists and community sociologists, have studied methods of community solidarity, cohesion, and celebration (Betancur, 2011; DePalma, 2009; Kezer, 2009; Sanjek, 2014). Business researchers emphasize a variety of aspects related to celebrations, such as where booths should be located during a festival to achieve maximum patronage (Gonzalez, 2013).

In contrast, community practitioners can be instrumental in brokering or facilitating a community's consideration of celebratory events. They can help communities locate local talent, navigate local ordinances and permits, obtain funding, evaluate outcomes, and assist with publicity and public relations. Community practitioners can also be cultural brokers, helping to break down stereotypes that lead to communities feeling misunderstood (Salazar, 2012; M.Y. Wu, 2014). The role of community practitioner should not be conceptualized as a leadership position, however, because that function must be carried out by local leaders, although a facilitator, or possibly, co-facilitator role is possible. A capacity enhancement paradigm identifies and engages indigenous leaders to enhance their competencies. How community practice evolves depends on local assets, circumstances, and needs.

Community Practice and Its Potential for Urban Communities

The voluminous literature on community celebratory events has significant gaps, and most notably, there is the absence of a guiding paradigm that stresses enhancing the capacity of a community sponsoring these events. This absence of a guiding paradigm, with its philosophical, theoretical, and value underpinnings, has limited how different disciplines can come together in service to communities rather than in service only to distinct subsectors of the communities. Communities consist of many different segments, which often coexist in the same physical space with but rare opportunities to come together. Community practice has evolved over the past three decades to take into account new paradigms and methods, and in the process of this evolution has embraced a variety of views on what constitutes practice, with community economic development being one of the most prevalent (Brueggemann, 2013; Weil, Gamble, & Ohmer, 2013). One perspective, or paradigm, is new in the past 15 years and unique because of its embrace of community assets (Delgado & Humm-Delgado, 2013).

Kretzmann and McKnight (1993) are widely considered early pioneers of community assets and capacity enhancement approaches through their work at Northwestern University's Asset-Based Community Development Institute. Considerable progress has occurred since that early period, creating new ways of conceptualizing community assets, particularly, but not only, in marginalized urban communities.

A number of scholars have paved the road toward the widening embrace of this paradigm, building on the groundbreaking work of Kretzmann and McKnight (1993). The late 1990s witnessed the first two scholarly books specifically on community capacity enhancement from a major academic publisher with the appearance of Oxford University Press's *Social Work Practice in Nontraditional Urban Settings* (Delgado, 1999a) and *Community Social Work Practice in an Urban Setting: The Potential of a Capacity-Enhancement Perspective* (Delgado, 1999b).

These two books, and other scholarly publications, have legitimized urban community practice, which may involve, for example, painting murals, planting community gardens, building playgrounds, creating community-built sculptures, and developing collaborative relationships with local establishments such as beauty parlors, bars, liquor stores, houses of worship, and grocery stores, as well as other, nontraditional urban enterprises. Urban assets have been conceptually developed, drawing on empirical evidence and validating the essential role that assets play in marginalized communities and how community social work practitioners can use them to enhance community well-being.

Celebratory events can potentially achieve many different instrumental (concrete) and expressive (psychological and social) goals, including giving neighborhoods a better brand, with a social identity that can attract social, political, and economic resources (Margry & Sanchez-Carretero, 2011; Roemer, 2007; Stodolska et al., 2011). Community celebrations can provide important insights that can be used to shape community practice by highlighting the types of events favored by a community and the success and failure of different types of events in the past.

Celebratory events serve socially important functions in shaping the perceptions and expectations of a community, and in mobilizing internal and external resources to, for example, convert dysfunctional public spaces and places into ones that are productive and inclusive (Ansari, Munir, & Gregg, 2012; Koch & Latham, 2012). Community celebratory events are being staged within the context of a distinct trend toward privatizing urban spaces, which curtails the events' potential to aid communities in their struggles to achieve social justice (Godfrey & Arguinzoni, 2012; Hogan et al., 2012; Langstraat & Van Melik, 2013; MacLeod & Johnstone, 2012). However, introducing festival crowds to urban spaces can help counter this trend by making it more difficult for politicians to sell spaces where there is a visible community constituency that would fight against such disempowerment (Jaguaribe, 2013a).

The *branding* of a neighborhood helps residents craft a narrative that shapes their identity and mobilizes their internal resources in service to their communities (Wherry, 2011, 95): "Ultimately, public opinion about the inhabitants of a place will determine what messages about the place resonate with mainstream audiences." However, if identity is imposed on residents it can have negative consequences for how communities are viewed, both externally and internally (Fainstein, 2010; Soja, 2010). Residents of a neighborhood or community may internalize a negative self-image, too, with dire consequences for current and future generations.

There is a distinct trend toward using tourism and cultural capital to achieve urban community development through branding (Clifton, O'Sullivan, & Pickernell, 2012; Iorio & Wall, 2012; Yao & Heng, 2013). Community development and celebratory events can unfold along a variety of dimensions (Derrett, 2004, 33): "The community development perspective on event[s] ... acknowledges the elements of community spirit and pride, cooperation, leadership, enhancement of cultural traditions, capacity to control development, improvements to social and health amenities and environmental quality." Culture as capital is closely associated with community cohesion, identity, and pride.

Community celebratory events have increased in popularity as a development strategy, and there has been corresponding clarity as to what these events signify and the increased need for research about them (Mair & Whitford, 2013). However, cultural events tied to community development goals must be authentic and enjoy the support of the community (Disegna, Brida, & Osti, 2011; Huang & Hsu, 2011); otherwise, they represent yet another example of cultural exploitation, commercialization, and marginalization (Forsyth, 2012; Spracklen, Richter, & Spracklen, 2013).

Corporate sponsorships over the past two decades have evolved and as sponsorship amounts have grown, so has accountability regarding how funds are utilized and how sponsoring corporations benefit from backing these events, as well as sponsors' expectations (Masterman, 2004, 261): "Essentially, sponsors have objectives that fit into one or more of these areas: to increase product or corporate awareness, to develop product or corporate image, to drive sales, or to develop market position. In other words, they seek a return on their investment. This shows how sponsorship has moved on from the days when sponsorship decisions were more philanthropic than strategic." This evolution is expected to continue in the future (Shah, 2014).

Economic exploitation of cultural symbols is certainly not new nor restricted to celebratory events, but it is quite pervasive in major community events as evidenced by how corporate sponsorships, and their symbols, find their way into commercials and advertisements enticing patronage from targeted groups (Fairer-Wessels & Malherbe, 2012; Finkel, 2010). However, the use of cultural symbols and advertising must be authentic or organizers and sponsors risk alienating their audience. For example, Oldenburg (1999, 103), in his classic book *The Great Good Place*, describes a southern-US German sausage festival that would be particularly unappealing from a food, music, drink, and decorative perspective; contrary to this dismal description, however, the festival attracts large crowds and is considered a huge success:

> The assessment of this sausage festival would be misleading if I failed to point out that it continues to be a success in terms of its repeated ability to draw crowds. Why? Several factors seem to account for the unmerited popularity of the festival. Most of the visitors, and particularly those under fifty years of age, have only those powers of discernment that experience has

provided. Bluntly put, they have not witnessed better community festivals organized at the grass-roots level. Parking is free and there is no staggering admittance charge, as confronts the visitor to the theme parks, World's Fair or Disney Kingdoms. Many undoubtedly find the event a welcome contrast to the slickness of the corporately-managed theme parks in which people are moved, stacked, and set in line with all due efficiency.

Oldenburg raises the importance of festivals and other celebratory events and how they must not value economic gain over the other elements associated with attendance. A trend toward commercialism and tourism as a major force in community development, including community branding, has the potential to subvert the role that celebratory events can play in enhancing community capacity and community identity (Ashworth & Page, 2011; Loukaitou-Sideris & Ehrenfeucht, 2009).

García (2004) addresses celebratory events as a major catalyst for participant engagement and urban regeneration and as a promising arena for community practitioners. Tallon (2014) and other scholarly publications have set the stage for advances in community capacity enhancement and expansion of urban community practice to include celebratory events (Altschuld, 2015; Delgado, 2015; Liberato et al., 2011; Littrell et al., 2013).

An assets perspective highlights a variety of key community institutions and public spaces that have largely been ignored by social scientists, such as small businesses, parks, museums, community gardens, murals, sculptures, and mutual aid societies (Allen et al., 2012; Eizenberg, 2012; Ore, 2011; Wronska-Friend, 2012). Cultures get manifested in urban spaces and places in artistic and non-artistic ways, influencing how communities express themselves (Baron-Yelles & Clave, 2014; Villa, 2000). Art can facilitate the development of an inclusive climate with transformative potential (Donish, 2013; Sonn & Quayle, 2013).

Grassroots memorials are another example of where public space and place are transformed (Langegger, 2013). Hawdon and Ryan (2011) address how community tragedies can use solidarity-producing events to help them cope. Memorial events are cultural artifacts and collective rituals that bring people together (Delgado, 2003; Hass, 1998). There are significant parallels between such public expressions and community celebratory events, including the use of prominent cultural symbols, since both give shape to public space for collective expressions and counter urban privatization trends (Margry & Sanchez-Carretero, 2011).

The importance of these institutions and spaces varies across communities and depends on local factors, including cultural values and the experiences of specific groups. Urban places belong to those who live within these spaces, including children and youth and people with differing cognitive and physical abilities (Kallio & Hakti, 2011; Sepe, 2013), and their needs and experiences should be taken into consideration.

In discussing Colin Ward's (1978) book *The Child in the City*, Banerjee, Uhm, and Bahl (2014, 123), emphasize age in understanding world views and experiences: "What differentiates a child's experience of space is scale, their attention to details albeit mundane from an adult perspective, their vivid and varied experiences obtained from previous mental associations or memories, and their perceptions shaped by more tactile instead of visual qualities of the surrounding environment. Ward contends the need for a shared city – that is, a city for children and adults alike – where the needs of the children have to be designed and shaped not as separate areas but integrated for their use anywhere and everywhere." A cookie-cutter approach is not recommended because local circumstances must themselves determine the best approaches (Altschuld, 2015; Delgado & Humm-Delgado, 2013).

The era of "urban renaissance" has ushered in exciting and innovative ways of thinking about community social work practice and contextualizing how capacity enhancement unfolds, particularly when embracing culture as an asset. However, there remain concerns about what this means for marginalized groups (Klemek, 2011; Magazine, 2011; O'Donnell & Tharp, 2012; Porter & Shaw, 2013). The goal of capacity enhancement cannot be achieved if only a select segment of the community reaps the benefits (Mukwada & Dhlamini, 2012).

One area that has not received due attention is how urban communities, with all of the conventional markers of distressed areas, create and manage to celebrate life in a culturally affirming manner. The role of fairs, festivals, and parades illustrates the concept of celebratory events, although celebrations are not restricted to these three types.

Conceptual Framework: The Community's Capital/Assets

Conceptual frameworks guide community practice assessments and interventions by outlining stages, activities, and major sociopolitical (interactional) and theoretical (analytical) considerations to develop an understanding of how to achieve the identified goals. A framework is a tool that can be used to build a project, but it is only as good as the practitioner using it. A capital/assets framework can guide a community capacity enhancement intervention and is highly relevant to celebratory events (Dunn-Young, 2012).

The concept of *community capital* captures a wide array of assets, including those addressed in this book as well as other less-known types such as "well-being," "welfare," and "personal" capital (Zahra & McGehee, 2013). This capital categorization can be more encompassing and attractive from a community practice perspective. Collapsing different forms of capital limits our conceptual understanding of how specific forms of assets can be enhanced and how they influence a community's well-being.

Community assets are socially constructed, embedded with values and meanings, and dependent on local circumstances (social, economic, political, and cultural). Assets can involve individual, group, environmental, and "invisible" forms. Assets, or capital, can be enhanced or diminished in importance depending on situational factors, do not have fixed values, and can continuously change as they interact with each other (Moeran & Pedersen, 2011).

Celebrations are indicative of a community's resilience and can be markers of a community's assets. The concept of *resilience* when it involves communities and celebratory events necessitates close examination, with culture being a key factor in any discussion of individual, organization, and community resilience (Bottrell, 2009; Wexler, 2014). Amundsen (2012) identifies the following six critical dimensions related to community resilience: resources, social networks, institutions and services, people-places connections, active agents, and learning. Each of these dimensions introduces a set of considerations that can enhance or diminish a community's resilience. Further, community resilience is not static and is subject to global challenges (Hopkins, 2011).

Although there has been a heavy emphasis on economic capital (generating money) because of the importance of tourism, taking a broader capital perspective on community events provides a more encompassing and comprehensive understanding of their importance, highlighting why the subject has relevance for helping professions and the social sciences, with particular appeal to those interested in community practice and research.

A socioecological view of capital, or assets, uncovers obvious and hidden values, illustrating how they are interconnected, and this broadens our conceptualization in order to tap them (Flint, 2010; Kimmel et al., 2012; Okano & Samson, 2010). Music festivals are a case in point (Oakes & Warnaby, 2011), as they take different forms based on local cultural considerations, challenging researchers to uncover how best to determine their impact on a community (Andersson, Armbrecht, & Lundberg, 2012). Art festivals face similar challenges (Grodach, 2010). Our ability to embrace assets will uncover numerous rewards but not without theoretical and practical challenges (Gibson et al., 2011).

There is an abundant literature on a capital (asset) construct, with social, economic, human, and cultural capital receiving the most scholarly attention (Delgado & Humm-Delgado, 2013). There is, however, a blurring of boundaries and use of different terms, depending on how practitioners or academics emphasize various dimensions. Communities may not care how such capital/assets get defined. Academics, however, do care. Having to convert local terms, particularly when culturally based and experiential, into academic terms places practitioners and academics into translator and interpreter roles that may be new to them.

The following seven types of capital/assets will be used in this book: social, economic, human, physical, political, cultural, and intangible. Each has distinctive qualities. However, the boundaries between types of capitals are permeable and allow flexibility in how they are operationalized, making them highly dependent on local factors to dictate their ultimate manifestation. The most successful celebratory events encompass all or most of these types of capital. Practitioners do not have the luxury of ignoring various forms of capital because all are needed when addressing, for example, marginalized communities.

Understanding community context (place and space) requires a broad appreciation of varied forms of capital (American Planning Association, 2011): "Helping a community to understand its historic, cultural, economic, and social context is an essential foundation for developing and building sense of place. This context includes a variety of community characteristics: population, demographic, and linguistic characteristics; physical and natural resources; cultural history; climate; customs; landscape features; design and architectural elements; local educational institutions; and temporary artistic and cultural exhibits, events, and spaces. A comprehensive reading or inventory of place can help a community begin to develop a voice for its narrative." Community assessment is only as good as the time and thought that goes into its conceptualization and the degree of community participation in the definitional process (Delgado & Humm-Delgado, 2013).

The economic value associated with celebratory events has dimensions that can often be considered *cultural capital* (Whitford & Ruhanen, 2013). An urban arts festival illustrates the multifaceted dimensions of cultural and economic capital (Quinn, 2005). Such festivals provide communities venues (safe space and place) to show off their cultural talent (Grodach, 2010). These festivals also provide a venue for generating *economic capital* through the sale of art and food to patrons (Pochettino et al., 2012). Festivals targeting ethnic and racial groups will rely heavily on the traditional foods of that community (Azar et al., 2013). These festivals have been marketed as community economic development opportunities, for example, in the case of Spain's Festival Internacional de Cine de Valdivia (Devesa et al., 2012).

Human capital, in turn, can be developed if there is an effort to enhance resident capacities and teach the next generation of residents how these events are planned, funded, and carried out, including through generating political support. *Political capital* is generated internally and externally. Principles of participatory democracy and volunteering opportunities (civic engagement) are created, garnering widespread political visibility, as evidenced, for example, by the number of elected officials wishing to have high-profile roles in parades. Coordinating or planning committees are developed, and they exercise various forms of political power in negotiations with local authorities and community stakeholders (Andersson & Getz, 2008; Kostopoulou, Vagionis, & Kourkouridis 2013; Myers, Budruk, & Andereck, 2010; Simon, 2010). *Social capital* is generated through creating new relationships or cementing existing ones.

Social capital and community events have benefited from attention in the literature, highlighting how celebratory events generate good feelings, relationships and partnerships, and connectedness and cohesiveness, although questions have been raised as to how these events bridge barriers between those groups that do not share similarities (Ansari, Munir, & Gregg, 2012; Wilks, 2011). Social capital has a prominent place in celebratory events, and will often be found alongside economic and cultural capital.

Social capital brings a relationship element into these events, particularly for residents who are disconnected from daily community life or, in the case of newcomers, missing their original homelands (Menjívar, 2006; Molitor et al., 2011; Whitford & Ruhanen, 2013). Rihova's (2013) research on festival attendee-to-attendee value co-creation highlights social capital (belonging, bonding, detaching, communing, connecting, and amiability) in situations that are socially dense. Socially dense events, or events with a large crowd presence, have a multiplier effect that can create feelings of warmth and belonging (Tovar et al., 2011).

Physical capital may be presented in vacant lands (Bowman & Pagano, 2010), capture buildings (Haines, 2009; Mallach, 2010; Skerratt & Hall,

2011), and other structures, and is produced when structures are created to house or host events, with these sites resulting in positive connections within the community (place and space). These structures can remain and, after the event, may serve other functions within the community. Plazas and public squares serve multiple social purposes and are places and spaces for gatherings and various types of celebrations because of their physical centrality and historical roots (Davis, 2001; Delgado, 2012; Steffensmeier, 2010).

Low (2000, 44) captures the importance of urban physical spaces, in this case the role of the plaza in Latin America, and why celebrations take on added meaning in communities: "Urban public places are expressions of human endeavors, artefacts of the social world are accommodated, and interpreted in the confines of the designed environment. Yet, as an object of study they have been relatively neglected ... partially because their social messages are so complex, and partially because the theory and methodologies for making these messages explicit were not available."

Physical places may be imbued with profound cultural history and social meaning, increasing their significance as host sites for celebratory events (Abidin, Usman, & Tahir, 2010). Public places with histories of activism (Arora, 2015; Gilster, 2014) and collective memories of social justice bring added meaning to celebrations staged there. Contested generational memories bring important interrelationship dynamics that can be reconciled or mitigated through an embrace of community events (Lacroix & Fiddian-Qasmiyeh, 2013).

Najafi and Shariff (2011) describe concepts such as *place attachment*, *place identity*, and *sense of place* which have emerged to capture the relationships people form with their physical surroundings. Sense of place, for example, helps us develop our understanding of how bonding and attachment to place emerges and shapes our impressions of particular places, helping us to assign meaning and values to a location.

Celebrations and place share a close relationship and play a powerful role in creating individual and collective memories and shared history with family and others. Positively altering a community's physical space is a key element in enhancing community capacity (Delgado, 1999b). Developing cultural districts is one way of designating a place and space within the community, with a specific purpose of creating an ambiance (space) conducive for community storytelling (Markusen & Gadwa, 2010).

Rosenstein (2011) makes five recommendations that cities can embrace to enhance neighborhood cultural life: cultural development must actively support diversity; cities must have a designated institution that supports neighborhood cultural activities and that identifies, assesses, and invests in cultural assets; the process for licensing cultural activities must be transparent and accountable to the community; cultural institutions and agencies must have an official role in helping to shape policies related to cultural activities; and, in some cases, a central authority should be identified for cultural activities.

Kim, Ahn, and Wicks (2014) identify three key attributes that affect the satisfaction of festival attendees: program content, program operation and organization, and advertising/public relations. This knowledge can help community social work practitioners establish the value of cultural products and differentiate their events, increasing the chances of achieving a competitive advantage when there are competing festivals occurring simultaneously.

Finally, *intangible or invisible capital* can be associated with events through timing, providing communities with a date when they can look forward to coming together, and creating positive identity (Stevens & Shin, 2012). Intangible capital can create a climate conducive to accepting new cultural traditions and views (del Barrio, Devesa, & Herrero, 2012).

Numerous services that are connected with artifacts are embedded with cultural value and serve as economic value and cultural value links (Throsby, 1999). As shown through a study of Chinese heritage assets in Bendigo, Australia, heritage assets can be integrated into interpretive themes or possibly anchored within communities to serve as cultural attractions with economic benefits, attracting visitors and tourists (Laing et al., 2014). The Edinburgh Festival, as another example, was created to allow the city to develop an international reputation for the performing arts, thereby engendering a creative identity (branding) with economic benefits (Prentice & Andersen, 2003).

Festivals generate human capital when providing residents with learning opportunities that can be transferred to other spheres in their lives (Whitford & Ruhanen, 2013). They can provide opportunities, for example, for youth to learn to play musical instruments and to perform at celebratory events (Conklin-Ginop et al., 2011). Youth participation translates into an appreciation of the meaning and value of civic engagement (Solís, Fernández, & Alcalá, 2013).

Developing a deeper understanding of the expectations, perceptions, and experiences of all age groups enriches our understanding of a community's celebratory events (Banerjee, Uhm, & Bahl, 2014), and how different groups can be engaged as attendees and participants, acquiring knowledge and competencies in the process, with a potential of achieving transformative goals (Mackellar, 2013b; Tisdall, 2013). Intergenerational learning is enhanced when youth and adults come together sharing goals and celebrations (Fitzpatrick, 2013).

Challenges for Enhancing Community Capacity

Community engagement has tremendous potential for social interventions by tapping assets in a manner that encourages democratic principles. This is an affirming and enhancing approach. There are challenges regarding how to define these assets since local sociocultural and sociohistorical circumstances wield prodigious influence over how these forms of capital get defined and prioritized by communities. It is necessary to point out that community capital is not static, and can increase or decrease in significance or, when faced with economic or natural disasters, disappear altogether.

This dynamic quality of human capital makes it challenging to assess and tap community assets (Beckley et al., 2008, 63): "Community capacity does not simply happen. Rather, it is developed and formed, or diminished and lost through response to changing conditions. Observable community capacity becomes manifest when there is a reason to act or to react. These reasons, or catalysts for action, may be positive or negative and we therefore describe them as opportunities and threats." This dynamic translates into opportunities opening and then closing, depending on local circumstances, with implications for staging celebratory events.

Community assets may not enjoy universal appeal across an entire community, as communities are not necessarily homogeneous in composition (Koutrolikou, 2012). A religious festival may attract segments of a community not affiliated with that religion; conversely, it may also serve to repel groups. Pride parades may be important manifestations of assets for the community but seen as detrimental by those who view LGBTQ people as deviants and a marker of community and societal disintegration.

These challenges are not insurmountable and are part of the lived experience. However, they do require serious thought to avoid unexpected setbacks in staging celebratory events. A failed event can cause incalculable harm to a community's image by publicly exposing its vulnerabilities to the external community, and possibly even further fracturing fragile relations between groups within the community. Celebratory events are not a "quick fix" to a community's concern but they can be transformative and can positively influence the identities of those involved and the places where they are held (Huang, Li, & Cai, 2010; Jaeger & Mykletun, 2013).

Goals of This Book

The following six goals will be addressed in this book:

1 Ground celebratory events within a community capacity enhancement paradigm to enhance appreciation of the historical roots and potential of fairs, festivals, and parades to transform urban communities.
2 Provide an in-depth understanding and appreciation of how celebratory events – specifically fairs, festival, and parades – can play instrumental and expressive roles in helping marginalized communities create and/or enhance positive identities.
3 Identify how multiple forms of capital are enhanced through community celebratory events.
4 Utilize case examples of fairs, festivals, and parades to integrate theory and research data to document their effectiveness.
5 Critique the advantages and disadvantages of using celebratory events as mechanisms for enhancing community capital.
6 Identify future directions that community practice embracing capacity enhancement can take to foster celebratory events.

The intent of this book is not to engender community practitioners who specialize as event planners. Rather, it is to increase their knowledge, skill, and comfort levels, and to demystify the use of these events to address a range of issues, as part of community practice.

Conclusion

This chapter provides a road map and addresses several key questions regarding why there is a need for a book on urban community celebratory events. We are surrounded by such events and much binds them together. Yet each type brings a unique set of historical circumstances that makes them important in community life. Celebratory events are not a panacea for marginalized communities. Yet they can be a part of the solution, having lasting benefits beyond an event's life span.

2 An Overview of Community Celebratory Events

Community celebratory events can be an all-encompassing term, and there is tremendous potential for rewards and challenges when incorporating them within urban social work practice (Devine, Moss, & Walmsley, 2014). It is important to obtain an overview of celebratory events before focusing on three specific types to be discussed here – fairs, festivals, and parades – and what makes them similar and unique. Further, events cover a wide variety of types, budgets, themes, geographical locations, and sizes, and share commonalities from having a community focus, crowds, celebration, to representing a concerted effort to involve residents in a planned activity, and more.

Celebratory events provide patrons, participants, and attendees of many different backgrounds with a range of activities and services, including, of course, food (Getz & Robinson, 2014). Fields and Stansbie (2003, 172) address this point: "Food, beverage, and celebration are inextricably connected. From social lifecycle events to giant hallmark events such as the Olympic Games or Super Bowl, the relationship between food and frivolity has been a close one. This is not to suggest that it is not serious business as well." The combination of food, frivolity, *and* business make events unlike any other business within a community.

The similarities between events do not take away from the unique dimensions and factors of community celebrations, including, for example, the central role that religion can play in shaping celebrations (Orsi, 2010). Parades, for example, involve crowds but because they are spread out over a long distance rather than concentrated within a narrow public space, they bring a different set of dynamics involving how attendees interact with each other, including the influence of noise on interactions (Farina, 2014). Festivals can go on throughout an entire day and even over multiple days and nights, whereby parades are time-limited to hours (Evans, 2012).

This chapter provides a variety of definitions of community celebratory events, taking into account how academic disciplines emphasize or de-emphasize various aspects and dimensions. This chapter also addresses the rewards and challenges of celebrations and their sociocultural origins, symbols, artifacts, and manifestations (O'Toole, 2011; Robertson, Rogers, & Leask, 2009). No community event is perfect, necessitating an awareness of their disadvantages and challenges, too.

Section 2 (chapters 5, 6, and 7) provides definitions of fairs, festivals, and parades, identifying major similarities and significant differences because of the uniqueness of each type of event, and how these events rely on themes, spaces, crowds, and places. These events can be viewed as separate entities; however, it is not uncommon to find them in various permutations and combinations with each other, for example, a festival can occur at the end of a parade (Davis, 2009).

Definitions of Community Celebratory Events

A search of the community celebratory events literature requires that a wide net be cast because there are various broad terms that have been used in relation to fairs, festivals, parades, and other celebratory events (Raj, Walters, & Rashid, 2013). The breadth of this field is exciting but terribly frustrating, too, since there are no simple answers for academics and practitioners to obtain when examining the influence of events on community life (O'Toole, 2011).

The following five overarching terms stand out beyond the generic concept of community celebratory events: *small-scale events* (Batty, Desyllas, & Duxbury, 2003a; Damm, 2011; Molitor et al., 2011); *field-configuring events*, which usually involve the arts (Delacour & Leca, 2011); *special events* (Ayob, Wahid, & Omar, 2013; Pereira, Rodrigues, & Ben-Akiva, 2015); *hallmark events* (Lever, 2013); and *creative and competitive events* (Moeran & Pederson, 2011). Each of these categories includes events of various sizes, complexities, and costs, complicating efforts to obtain an inventory, unifying definition, and universal understanding of celebratory events (O'Toole, 2011).

Definitions play an important part in bringing to fruition community-focused or community-sponsored events, and that it is why there is a need for boundaries as to what constitutes community celebratory events. Readers may be in a more propitious position to embrace an existing definition or develop one that best meets their own needs. Regardless, a definition is in order.

Celebratory events have been defined in a variety of ways, and these definitions influence how events unfold and are understood (Mangia et al., 2011). Loukaitou-Sideris and Ehrenfeucht (2009, 63) approach celebratory events from the perspective of physical place, addressing how city streets get converted into welcoming spaces to hold celebrations: "Parades can be seen as territorial acts. Their participants temporarily claim the public space of streets and sidewalks and stir local and sometimes even global audiences. Participants appropriate the pavement and 'enact the streets' for their own performances of ethnic, sexual, political, or religious identities." This perspective broadens what, and how, urban space can be used to host a celebratory event.

Streets and sidewalks can be transformed from drab public spaces consisting of nonharmonious sounds to ones that are colorful and filled with music and laughter, with all kinds of foot traffic. These public spaces can also include alleyways, which are rarely discussed in the literature (Seymour et al., 2010; Wolch et al., 2010), although alleys are common public areas in cities.

School grounds can become sites of events, such as student parades and music concerts (Steele, 2013). Parks and community gardens, too, can be transformed. Urban green spaces can be tapped to host events and capture socioecological memories (Barthel et al., 2014; Baycan & Nijkamp, 2012). For this to occur, these spaces must be safe and accessible (perceived as welcoming) to attendees and event workers (Platt, 2012).

Definitions ground community practice by providing boundaries and explaining what activities and tasks are to be included or excluded when considering events. Celebratory events are predicated on clear boundaries, geographical and psychological, including expected behavior for those attending and hosting the events. Definitions assist communities in determining how celebratory events have meaning to those hosting them, with these meanings typically clustered into various categories.

The following six definitions provide an overview of how celebratory events have been defined in the professional literature, although this list is far from exhaustive. Yet they capture themes that will permeate throughout the discussions on fairs, festivals, and parades in this book and manifest themselves in a variety of ways, highlighting the multitude of ways that these events can be found in urban communities.

Community events, according to Jakob (2013, 448), can be defined as "the deliberate organization of a heightened time and place." Goldblatt

(2002, 6) says that an event is "a unique moment in time celebrated with ceremony and ritual to satisfy specific needs." Getz (2012, 40) provides a broader definition: "Planned events are live, social events created to achieve specific outcomes, including those related to business, the economy, culture, society and environment." Carlsen (2004, 247) submits that an event "involves some form of celebration or occasion which is carefully planned to meet specific or economic objectives." J.P. Smith (1986, 4) defines events as "unique, infrequent, short-term activities which depart from everyday life and involve the entire community. They range in size from the small children's magic show put on at a local park for a small local crowd, all the way to a major festival that attracts thousands of visitors and takes place over a course of a week." Jackson's (2008, 161) definition is more inclusive of many factors and resembles the definition guiding this book: "Community festivals and celebratory events acknowledge people, places and anniversaries and provide sites for social interaction, community participation, and expression of group identity. They act as artefacts of local concerns and interests and as a means of achieving political and economic ambitions and as a vehicle for meeting social and community needs." Geographical event boundaries, and more specifically those of fairs and festivals, facilitate the mobilization and concentration of resources in a defined space (Lampel, 2011; Moeran & Pedersen, 2011).

The above definitions purposefully are short and stress event experience, various lengths of duration, space, place, themes, and the achievement of multiple goals. Most definitions of celebratory events have in common an embrace of culture, crowds, origin within the community, and the involvement of ceremony and ritual (Brown & James, 2004). Events, regardless of type, share a core set of activities, have a theme, occupy a specific location and venue, and take place at a specific time period; they might have an admission fee or be free to attend (Salem, Jones, & Morgan, 2004).

A *celebratory event* for our purposes can be defined as an indigenously planned activity that can be educational, civic, cultural, and/or public in nature; places emphasis on crowd engagement; and conveys a narrative that has community significance, enhancing local competencies in the process. This definition grounds events within a community context, stressing engagement and multifaceted goals.

The size, focus, goals, crowd expectations, time span, and activities related to events vary according to local circumstances, yet there is a common base uniting them. This perspective does not refer to mega-celebratory events such as the Olympics (Gilmore, 2014; Sadd, 2012; Shin, 2012) or large-scale exhibitions (Ballester, 2014; Müller, 2014), which are not neighborhood-centered and which, incidentally, can result in major negative community transformations caused by massive displacements (Caramellino, De Magistris, & Deambrosis, 2011; Greene, 2003).

An encompassing definition facilitates creativity and inclusivity, qualities needed in community practice. However, such a definition may be too broad, translating into permeable boundaries that result in events meaning all things to all people. A very narrow and highly detailed definition makes it easier to determine whether an event "qualifies" as community and celebratory. Domains capture key elements of these events in similar fashion to how Schippers and Bartleet (2013) developed their nine domains for classifying community music and Lekies' (2009) classification of youth community events. Domains provide specific boundaries that are sufficiently broad to encompass key elements.

Schippers and Bartleet's (2013) nine domains or categories for classifying community music serves as an excellent example of such an effort: infrastructure, organization, visibility/public relations, relationship to place, social engagement, support and networking, dynamic music making, engaging pedagogy, and links to school. Each domain highlights unique aspects that can be appreciated by themselves or in combination with various types (permutations and combinations). Lekies (2009), in a novel approach, undertook a content analysis of a small US town's newspaper over a one-year period, and identified 12 categories of youth community engagement events: "programs, clubs and special events; fundraising and community service; business and community support; participation in community events; school events; athletic and other performances; employment; involvement in local planning and decision making; serving as a community representative; visibility and recognition; criminal activity and accidents; and use of public space." The reader may not consider some of these community activities to be "events," celebratory or otherwise, highlighting the challenge of what is included or excluded in understanding this form of celebratory event.

The Purposes of Celebratory Events

Celebratory events help attendees break from the predictability of daily life (Mackellar, 2013b). However, van Heerden (2011, 55) questions this very assumption and uses the concept of liminality (the altering of the ordinary to the extraordinary) to argue a central point by identifying six key factors related to assessing the role and importance of suspending the ordinary: "extensive planning and preparation, different senses of time, the alteration of everyday routines, re-discovery and re-appropriation of private and public spaces, the activation of festival spaces, and the reworking of rules." Each of these dimensions of liminality transforms an ordinary moment into one that is extraordinary.

Celebratory events must fulfill a variety of purposes beyond celebrating a moment in a community's history, and these purposes can often address multiple local needs. Events are best when they address multiple aspects of a community internally and externally to fulfill their potential promise for transformation (American Planning Association, 2011): "an articulation of the historic, cultural, economic, and cultural context of the community; a commitment to the reinforcement and enhancement of the community's identity; and the implementation of policies, regulations, and incentives that support and enhance this evolving identity." These events can be simple or ambitious in scope and can be easily modified to take into account local socioecological circumstances and goals. For example, computer-simulated models have been developed to measure the impact of changing routes on parade crowds to minimize disruptions (Batty, Desyllas, & Duxbury, 2003a).

Celebratory events should be organic in character, originating and being fed by the community hosting the event. Outside support can be obtained, but decision-making control should rest within the community. Similar to those for community music events, definitions of celebratory events can be either too broad or too specific to be of use for practitioners (Leglar & Smith, 2010; Schippers & Bartleet, 2013). There is a general consensus that music festivals, regardless of type and location, can enhance various forms of capital such as social, economic, and cultural (Mair & Laing, 2012; McCarthy, 2013). However, studying such events and their impacts can be challenging as multiple academic disciplines engage in such studies, with each discipline bringing its distinct theoretical bias and language to the subject (Baum et al., 2013; Mangia et al., 2011). Practitioners need a Rosetta stone to help them translate across academic languages because the same phenomenon may be called by different, discipline-bound terms.

Arcodia and Whitford (2006) comment on the number of scholars that specifically focus their attention on ascertaining the economic value of festivals on host communities. This is both a strength and limitation in the existing literature. A festival's economic value can be overstated, or even overlooked from a local or regional perspective, or it may not be spread out evenly across the community (Kostopoulou & Kalogirou,

2011). This necessitates a direct and indirect conceptualization (Lee & Arcodia, 2011; Litvin, Pan, & Smith, 2013; Terry, Macy, & Owens, 2009), highlighting the challenges of measuring what many consider the easiest asset to measure. The impact on the informal economy has largely gone unnoticed, adding to the challenge of measuring and tapping economic capital benefits (Brown, Lyons, & Dankoco, 2010).

There is a tendency to reduce this capital to price as a true measure of economic value (Throsby, 1999). Cultural economists have recognized how this form of capital can be combined with other forms of capital, which can alter and even enhance economic capital. Participation in celebratory events, however, has been considered to be under-researched (Palma, Palma, & Aguado, 2013). Events must be viewed from a comprehensive perspective, increasing their value to a community, although this compounds how they get researched and evaluated (Brown & Trimboli, 2011).

Regardless of the shortcomings of focusing exclusively on the economic value of an event, however, there is no denying that economic capital a key asset that must be a part of any consideration of a festival or other celebratory event, particularly in communities that have severe economic hardships (Gibson et al., 2011; Yoon, Lee, & Lee, 2010). The economic motivation should be a part of other social, cultural, and political goals, facilitating a multifaceted social change agenda.

The festival literature highlights how cultural capital can lead to economic capital (Van Aalst & van Melik, 2012, 196): "Festivals in particular give strong impetus to the urban economy; they operate at the interface of art and culture, the media, tourism and recreation. Figuring prominently in the development and marketing plans of many cities, festivals are deemed to foster a positive image of destination ... Given their multiple meanings and functions, then, it is no surprise that festivals have increased enormously in number, diversity and popularity." Cultural capital can be manifested in a variety of ways and clustered, further reinforcing how this asset can be transformed, measured, and incorporated into interventions (Lyck, Long, & Griege, 2012; Madsen et al., 2014; Stern & Seifert, 2010).

Community celebratory events are not possible without collaborative partnerships and an infusion of economic funds (Liu & Lin, 2010). It is important to emphasize that community assets can be synergistic, with enhancement in one translating into enhancements of others, although not necessarily of equal strength or occurring simultaneously. This requires using a socioecological perspective and an interest in achieving a comprehensive community understanding.

There are numerous challenges in gathering economic capital data (formal and informal economy) which, if addressed with thoughtfulness and a multidisciplinary perspective, can guide the conceptualization, operationalization, and measurement of other forms of event capital (Loukaitou-Sideris & Soureli, 2012). Events represent opportunities for economic activity as street vendors, for example, among those who are undocumented and are limited as to where, and when, they might sell their foods or other items (Rosales, 2013). There is no reason why economic capital must be the exclusive domain of economists. Different disciplines bring added dimensions to this form of capital and facilitate its integration with other forms of capital, and in the process introduce innovative thinking about this asset, as evidenced by the high reliance on economic capital in celebratory events.

There is an absence of research on the social, cultural, and/or political significance of festivals and events in the life of urban groups (O'Sullivan, Pickernell, & Senyard, 2009), particularly regarding marginalized communities. The dearth of information on these less popular forms of capital and groups creates a distorted view of the value of celebratory events and undermines efforts at using them in community practice. Popular media coverage of community celebratory events tends to focus on crowds, activities, and the central reason for the celebration, such as religious, ethnic heritage, founding days, and championships. Ethnic media sources can play a significant role in shaping how these events are shared with the broader community, and communications with these media must be cultivated by event organizers (Lindgren, 2013; Ross, 2013), for example by having someone in charge of public relations and media contacts.

Moeran and Pedersen (2011, 3–4) address what they consider to be the absence of serious research on celebratory events, which limits our understanding of their importance: "In spite of their obviousness and the fact that they have been increasing in number very rapidly over the past two decades ... fairs and festivals have, until recently, largely been ignored by scholars working in such disciplinary fields as sociology, anthropology, strategy and management, as well as economics, although ... not by geographers or historians." Berridge (2010) addresses the subject of event pitching or the winning of the right to sponsor or host by an organization or group, another aspect which is often overlooked in the professional literature.

Mattivi and colleagues (2011) introduce the concept of sub-event recognition and its relationship with time and time constraints, and the importance of developing a sounder theoretical and empirical understanding. Events are held over a prescribed period of time and so does each stage, too, unfold over a time span. Invariably, there is an opening and closing ceremony to mark the beginning and end, with activities between these two periods occurring in precise order.

Celebratory events are complex in character, necessitating a wider and more nuanced lens for deepening our understanding of their values, or benefits, to communities and to society as a whole. To achieve this portrait is labor intensive and requires a multidisciplinary perspective, with community residents playing an active part in shaping our understandings, and a willingness on the part of evaluators to ask questions that normally would not fall under the purview of evaluation. For example, how does attending a community celebratory event alter attendees' views about the community?

Why Attend a Community Event?

The question of why one would attend a celebratory event may appear simplistic, but it gets at the root of motivation, expectations, satisfaction, and the main reason for holding an event in the first place (Yang, Gu, & Cen, 2011). There is no set number of reasons as to why we are motivated to hold, attend, or even work at a celebratory event. Reasons may cluster into categories, such as fun, as discussed later on in this section. However, answering this question without a nuanced understanding represents a serious loss to our appreciation of why to attend, or why to hold a community event (Ayob, Wahid, & Omar, 2013; Dawson & Jensen, 2011). Holding an event represents a strategic decision, with clear goals concerning the outcomes and justification for using community resources.

There is extensive literature on why to undertake a celebratory event from a stakeholder perspective, with certain stakeholder types being emphasized (Jakob, 2013; Schmallegger & Carson, 2010). Vested local event interests must be present for these events to enjoy community-wide appeal (Abdulla, 2013). Slater (2011) raises a cautionary point that events can be sponsored by stakeholders that should not be initiating these events because they lack community awareness or have highly personalized agendas. Events with highly politicized agendas represent political stakeholders' interests, compromising the potential goals and spirit of the events.

There is an absence of literature on why a local resident should attend community events, and with whom, creating a serious lacuna in our understanding of community celebratory events. One notable example is the limited research on how the broader communities feel they are a significant part (or have ownership) of these events, including attending and contributing to them (McGregor & Thompson-Fawcett, 2011; Rogers & Anastasiadou, 2011). This is a sad commentary given that celebratory events are about communities.

Are these events just for residents, visitors, or even former or returning residents, or all of the above? The answer goes a long way toward determining expectations. Celebrations can be a mechanism for welcoming back those who have left their communities through the creation of a social space that heightens a sense of community and belonging regardless of where attendees actually reside (Ziakas & Costa, 2010a). Former residents can remain attached to a community, even when significant time and distance has occurred, giving events a level of significance as a homecoming.

There is a strong relationship between festival attendance and emotional connection, and when this connection occurs it contributes to attendees developing a sense of community (Van Winkle & Woosnam, 2014; Van Winkle, Woosnam, & Mohammed, 2013). Family decision-making brings an added view to attendance motivation, which is often characterized as individually based (Kim et al., 2010). Attendees' motivation takes on added meaning when ethnic and racial groups, with strong family ties and traditions but otherwise limited in their mobility and acceptance outside of their immediate community, sponsor celebratory events. Events necessitate taking a group perspective. Bringing a group decision-making process to an attendance assessment adds a level of complexity at arriving at an answer.

There are many urban communities with histories of being a port of entry for newcomers who then disperse to other communities when they have adjusted to their new surroundings. Events provide former residents with an opportunity to reconnect and recapture memories that helped ground them on where they have been and where they want to go for themselves and their families.

Motivation for attending a community-sponsored event is a topic that has historical appeal and importance, particularly among tourism scholars interested in mechanisms of community development (Yolal, Cetinel, & Uysal, 2009). An event's *gravitational force*, which refers to its ability to draw audiences from a geographical distance, can enhance its economic impact by broadening the appeal of an event beyond a narrow geographical confine (Adams & Adams, 2012). Celebratory events have a central goal of achieving wide community appeal because of their economic, social, and political goals.

Crompton and McKay (1997) study festival attendance and find that visitor motivation can be classified into the following six categories: cultural exploration, novelty/regression, recovery of equilibrium, known group socialization, external interaction/socialization, and gregariousness. Each category taps a motivational dimension that can be enhanced with careful planning, but each consideration first must be firmly grounded within a local context to capture its cultural significance and manifestation.

Greater attention to the various motivations for attending events has led scholars to consider sociodemographic factors. Leonard and McKnight (2013) study adolescent perceptions of Belfast parades and find viewing attendance in celebratory events from an age perspective to be important. Age can also influence attendance at LGBTQ Pride (M. Brown, 2014) and music events (Packer & Ballantyne, 2011). Other demographic factors wield significant influence as well. Gender plays an influential role in music festivals (Blešić et al., 2013; Bowen & Daniels, 2005). Looking at repeat attendance (loyalty) at cultural festivals brings another dimension to our understanding of attendance, taking it away from a single-episode experience and toward the potential for creating a legacy for those with traditions of attending (Lee & Hsu, 2013; I.S. Lee et al., 2012).

Communities are not homogeneous: different segments of their population view celebratory events differently, complicating our understanding of events and what factors we use to determine their success or failure. Participatory activities, games, food, music, dancing, and general atmosphere will rarely hold the same appeal for all attendees – the appeal will differ based on sociodemographic factors, including religious background (Calestani, 2013; Stone & Millan, 2011). It becomes important to identify the demographic characteristics of attendees so that events can be structured to attract certain groups.

A gendered perspective raises issues related to male privilege and sexism as operating forces and to what extent gender influences stakeholder power and influence. Studies can look at the gender of celebratory event leaders and stakeholders, and how it influences the way events unfold. This perspective is sensitive but cannot be ignored if community practitioners are to understand the meaning of celebratory events (Phi, Dredge, & Whitford, 2014).

Sexual identity highlights other aspects related to social justice, identity politics, disenfranchisement, and empowerment, and how the age of participants and the private and public nature of their relationships influence their expectations and perceptions (Clarke, Burgoyne, & Burns, 2013). It, too, is a sensitive topic but cannot be ignored. For example, are younger LGBTQ participants more likely to question the value of pride events when compared with their older counterparts?

Miranda's (2010) case study of celebratory events and community gardens in a Puerto Rican community in the Lower East Side of Manhattan, New York City, focuses on the close relationship between capacity enhancement and a celebratory parade and festival that serves multiple goals, including demonstrating community power (political capital), enhancing cultural pride (cultural capital), generating economic outcomes (economic capital), encouraging use of community gardens (physical capital), and bringing together multiple generations of residents (bonding social capital).

Miranda's case study integrates festivals, parades, and fairs in one urban community, showing how multiple forms of capital interact with each other and influence celebrations (Cicea & Pirlogea, 2011). What may start off as a single event can evolve into an entire week devoted to celebrations, including multiple forms of events, challenging control of the evolution of the message because of the increased number of stakeholders.

The concept of *fun and play*, central to the lives of all communities, has generally escaped serious scholarly attention when discussing celebrations, although it can be a powerful motivator for attending (Bates & Fortner, 2013). Community events can be viewed from a wide range of perspectives, drawing multidisciplinary attention, including fun and play. Eberle (2014) introduces the role of fun and play as leisure and its evolution and various manifestations, showing how this concept can be applied in a variety of settings and activities. There are few activities that lend themselves to fun and play that can appeal to all age groups, but community celebratory events are a notable exception.

Landers (2012) introduces a geographical context and sees play as having a potential social justice foundation by focusing on the interactions between strangers and enhancing familiarity with people from different backgrounds (bonding and bridging social capital). Fun and play can be applied to community celebratory events to achieve new understandings on participation and the creation of long-lasting collective memory. Attendees at community events do not have to engage in activities to have fun; observing others can be fun, too.

Urban Social Anchors, Cracks, and Ambiance and Community Celebratory Events

The attractiveness of community celebratory events has led to the use of varied theoretical concepts, which deepen our understanding of how they have the potential to enrich communities and society as a whole and which facilitate academic collaboration. Two such concepts are *social anchors* and *urban cracks*. Each captures a view of events and informs how best to select and support these events.

An anchor is such a great metaphor. The concept has been applied to community organizations that fulfill a variety of important functions, including serving as a focal point (or anchor) of social networks, resulting in the enhancement of community identity, trust, and reciprocity among residents (Alonso & O'Shea, 2012; Clopton & Finch, 2011). In addition, these organizations can draw on their geographic location to facilitate the use of various community assets, whether they are physical or intangible. Urban social anchors can take many forms beyond stressing important institutions in community life. Community gardens and plazas are examples of how these physical spaces can bring communities together to share and connect through special events and daily life.

The concept of *social anchors* can be applied to collective memory, social relations, and social networks (Benski et al., 2013; Gongaware, 2010). Community celebratory events can be social anchors in fostering social capital. When these events are conceptualized as anchors they provide a visual means for capturing various forms of capital, by altering the landscape of a community by introducing new structures and color to what is often varying shades of gray (Alonso & O'Shea, 2012). Culture, too, can serve as an anchor for a group, institution, and community (Bressan & Alonso, 2013; Ghaziani & Baldassarri, 2011).

Verschelden and colleagues (2012, 278) put forth the concept of *urban cracks*, describing them as "as spatial, temporal and relational manifestations of social and political struggle in the city." Urban cracks, therefore, are urban spaces that facilitate or encourage dissent and thereby make consensus arduous to achieve. These cracks cause tensions and conflicts that can further damage community relations, but they can also translate into opportunities, particularly in regards to urban community events, which can facilitate collective learning and the exercise of participatory democracy, working against the urban cracks.

Thibaud's (2011) perspective on urban space brings an important dimension to how urban celebratory events transform communities.

Urban public places stimulate sensorias that are dramatically different from those of rural spaces (Thibaud, 2011, 43): "Urban space provides numerous ambiances to be felt with all of the senses. Whether we think of a lively outdoor marketplace or a dull parking lot, an attractive historical center or a casual subway station, the very way we relate to those places is based on the experience they involve. It is a matter of light and colour, sound, smell, touch and heat, and also the manner we walk and talk, move and look, relate and behave. Urban ambiances operate each time a subtle interweaving of synesthesia and kinesthesia occurs, as a complex mixture of sensation and movement." Similarly, celebratory events can alter physical environments in a way that can alter one's perception of the surroundings.

Urban spaces can offer opportunities to meet and interact with groups outside of our own. Peters and de Haan (2011) note that in general ethnic diversity can appear in public spaces as a result of newcomers and particularly in spaces that are ethnic-specific. These ethnic-specific places, however, can still result in interethnic contacts and relationships, which can lead to greater awareness and acceptance of diversity and the development of social capital.

Types of Community Celebratory Events

Celebratory event classification is a task that brings inherent challenges but is essential for developing an understanding and appreciation of fairs, festivals, and parades, and how they fit within the constellation of community events, including where they overlap with each other and where they make unique contributions (Tkaczynski & Rundle-Thiele, 2011).

According to Getz (2012, 41), a leading scholar on the subject of community celebratory events, there are six major types of events: "(1) *Cultural celebrations* (festivals, heritage, commemorations/carnivals, Mardi Gras/religious rites/pilgrimage/parades); (2) *Business and trade* (meetings, conventions/fairs, exhibitions/markets/corporate events/educational scientific congresses); (3) *Arts and entertainment* (scheduled concerts, shows, theatre/art exhibits/installations and temporary art/award ceremonies); (4) *Sport and recreation* (league play, one-off meets, tours/fun events/sport festivals); (5) *Political and state* (summits/royal spectacles, VIP visits/military (tattoos), political congress); and (6) *Private functions* (rites of passage/parties/reunions/weddings)." McGurgan, Robson, and Samenfink (2012) identify five different types of celebratory events which overlap with Getz's six types: hallmark or mega-events, fairs and festivals, sporting events, business events, and special events.

Community events can take different manifestations and have different goals. How and why these events take their particular form is influenced by local circumstances, goals, resources, and the competencies of in leadership and supporting roles, opening up endless ways that community practitioners and scholars can approach celebratory events. This flexibility makes it especially important to undertake a community assessment to determine the most appropriate form of event.

Rewards Associated with Community Celebratory Events

There are numerous instrumental and expressive rewards that can be gained through application of a community's capacity enhancement paradigm to fairs, festivals, and parades, justifying the expenditure of time and funds, and the seeking of community participation (Clifton, O'Sullivan, & Pickernell, 2012). These potential rewards can be maximized when community practice embraces celebratory events, including accessing previously untapped sources of potential funding. Community practitioners bring knowledge of funding sources and how to obtain them. This knowledge opens up new doors for community groups without experience seeking funding from corporations, foundations, and government sources. The bringing together of fun and fundraising has a long history in society, and community celebratory events represent but the latest efforts at combining these two (Tennant, 2013). Yet events can certainly go beyond fundraising.

The following eight benefits of community celebratory events illustrate the range of possible positives they offer for urban marginalized communities. Celebratory events help marginalized communities enhance their positive profile; identify and cultivate indigenous leadership; enhance community capital; mobilize internal and external resources for social change; assist in the development or enhancement of community identity; tap the assets of overlooked population groups; increase and disseminate lessons learned so that other communities can benefit; and generate funding with sufficient flexibility to support initiatives of the community's choosing.

These benefits may vary in intensity and importance depending on community circumstances, goals, and the extent of community participation. It is essential that an in-depth assessment focus on the particular benefits that are considered high priority to the community undertaking an event (Dunn-Young, 2012). Community events are best tailored to maximize the use of available formal and informal resources. These rewards may not exist in equal strength or in similar time periods; they do, however, illustrate the potential range of positive social change possible as a result of a community event. A significant investment of time and energy, as well as financial and other resources, including political will and community participation, is necessary to achieve these rewards and benefits (Litvin, 2013). This is a tall order, but essential.

Community Celebrations as Rituals and Authenticity

Community celebratory events can be viewed from multiple perspectives with some viewpoints having greater saliency and influence in shaping how these events are researched and practiced. Viewing events from a collective ritualistic perspective (as formalized gestures and procedures with profound social/cultural meaning) introduces an important cultural and contextual grounding (Collins, 2013; Pearlman, 2013), facilitating an understanding of how various forms of social support are inherently community celebratory rituals (Li, 2014; Roemer, 2007). Micro-enterprises – small business that are either run out of someone's home or as part of a larger business establishment – can participate in celebratory events and their involvement can be viewed from a ritualistic (Connolly, 2010) as well as economic perspective.

Durkheim (1915) put forth a conceptual argument that cast rituals as expressions and reinforcements of cognitive functions that serve to unite communities. By incorporating rituals, celebratory events, particularly those with heritage themes, can also create new meanings by encouraging new traditions that have particular local appeal. Lena's (2011) analysis grounds festivals within a sociocultural, sociohistorical context, making their significance more far-reaching than what is generally seen by community outsiders.

The study of rituals historically has attracted many different academic disciplines and has wide appeal across a number of arenas that can encompass political, economic, religious, and even ordinary lives (Wang, 2012). The acceptance of rituals as an analysis frame increases the saliency of this perspective, but it also brings challenges because of language differences on rituals resulting from different academic disciplines and the importance of emphasizing localized context to give meaning to rituals.

In a case study of the Oje'nmeho Cultural Festival of the Awume people of Nigeria, Idoko (2012) argues that sociopolitical, economic, and cultural perceptions of people can be appraised through participatory events, such as community theater. Events offer community residents an opportunity to share their thoughts and even engage in decision making, thus aiding community and societal transformation by emphasizing the principles and process of participatory democracy (Simon, 2010). Knottnerus (2010), in addressing the interrelationship between celebratory events, collective rituals, and emotions, argues that greater emotional connection to events translates into greater group commitment and solidarity.

Smith-Shank (2002) merges multigenerational benefits of community rituals and celebratory events. Community celebrations can be viewed as critical markers of rituals and in the process serve to ground individuals with a sense of belonging while also providing an opportunity to introduce a process of questioning. Thus, they represent an approach toward creating an understanding of history, communities, and identity. Celebrations also provide an opportunity for socialization to transpire among all age-groups, including children. Rituals are conducive to being practiced and understood individually and collectively, and they can only be properly understood and appreciated when contextually grounded.

In researching the Orange Order of Northern Ireland, Edwards and Knottnerus (2010) find that their rituals are highly ranked and symbolic practices, serving the members' beliefs and power needs, with a highly publicized parade being a mechanism for achieving these outcomes in public spaces. The Orange Parade, however, is not without its share of tension and conflicts, both political and religious (Jarman, 2003; Ryan, 2000). Kates and Belk (2001), in researching LGBT parades, refer to these celebratory events as "collective rituals of consumer resistance," uniting rituals and politics.

Community celebratory events have many key elements that are traditionally associated with rituals, such as important roles related to communication, learning, and cultural values, and they function to draw public attention to values, thoughts, and behaviors (Santino, 2011; Smith & Stewart, 2011). Not all urban public spaces can be considered ritual-conducive spaces and when celebrations are held in ritual spaces, events honor these spaces (Alem, Vaziri, & Sharif, 2014; Ashebir, 1999).

Iltis (2012, 17) identifies four ways that rituals reinforce social reality, with implications for celebratory events emphasizing cultural origins and heritage:

> First, by creating a social reality, rituals establish or reinforce expectations, relationships, and roles; they create a web of social bonds. Second, by

inviting participation in a social reality, rituals maintain social stability and harmony; they create sustaining social structures. Third, rituals by placing individuals within a social reality enable individuals to understand themselves as part of specific groups invested in particular activities, commitments, and traditions; rituals by creating social reality allow individuals to understand their position within the social geography of the world. Fourth, rituals by placing humans within a social reality disclose the significance and meaning of time, including the passages of human life, from reproduction, birth, marriage, and suffering to death.

Ilitis sheds light on the functions of rituals and their importance to communities, including celebrations. Understanding these rituals provides outsiders with a window to understand and appreciate celebratory events.

Authenticity implicitly permeates celebrations, making it important to give it the attention it deserves. The construct of authenticity is embedded with memory, place, identity, and cultural significance (Waterton & Smith, 2010). Authenticity takes on event significance because it is often a key element in the expectations of both attendees and those staging the event (Chhabra, 2010; Guttormsen & Fageraas, 2011; Yang & Wall, 2009). Authenticity can be conceptualized as the "glue" bringing a community together.

Castéran and Roederer (2013, 153) define *authenticity* as "a concept that encapsulates what is genuine, real, and/or true." Their review of the authenticity literature concludes that authenticity has significance for the discussion of tourism and celebratory events, although the concept has enjoyed considerable debate, not surprisingly, based on academic discipline, ideologies, and what constitute definitions of "sincerity," "purity," and "originality" (Alivizatou, 2012; Andriotis, 2011; Grazian, 2010; Lindholm, 2013; Yang & Hu, 2011). The term *organic* is closely associated with authenticity and refers to the process giving rise to ownership and decision making (Alonso & O'Shea, 2012).

Gundlach and Neville (2011) view authenticity from a continuum rather than a "yes" and "no" point of view as a means of dealing with the complexities and challenges of operationalizing this concept. Pine and Gilmore (2007), in turn, classify authenticity into one of five categories: original, natural, exceptional, referential, and influential. Heitmann (2011) grounds authenticity historically (Greco-Roman), identifying ways it has been applied.

Authenticity indicates a sense of true, sincere, or original elements, and its definition is greatly influenced by the context in which it is viewed, necessitating that this grounding be taken into account by event scholars, organizers, and evaluators (Theodossopoulos, 2013). Cohen and Cohen (2012) advocate for further exploration of authenticity by focusing on who is authorized to determine whether an event is authentic, uncovering issues related to power and contestation.

Authenticity is associated with a range of topics related to celebratory events, including urban regeneration (Kurdia, 2013; Ulldemolins, 2014); sustainable development (Simao & Partidario, 2012); branding (Bruhn et al., 2012); social marketing (Winter et al., 2013); place (Zukin, 2011); leadership (Freeman & Auster, 2011); and tourism (Croes, Lee, & Olson, 2013; Rickly-Boyd (2012). Carmody (2013) and Lindholm (2013) point out that existential authenticity, or activity-based experiences, has increased in prominence within the scholarly literature as a means of capturing and stressing the interrelationship between place and experience.

Stakeholders, particularly those tied to local governments, wield influence in undermining authenticity in the hopes of creating wider appeal for tourism (Slater, 2011). Defining what is meant by celebratory event authenticity rests with the community hosting the events (Croes, Lee, & Olson, 2013). Their perspective can easily get lost among academics debating the meaning and nuances of this concept.

Authenticity lends itself to community development by stressing the importance of meaning to residents (Onyx & Leonard, 2010) and how it shapes representation of the community (Kaufman, 2011). However, no community is homogeneous in composition and there will be varying differences according to subgroups. Identifying common core elements and reconciling differences becomes an important task in staging celebratory events that stress historical events and cultural heritage. Who determines authenticity and how they do so impacts the organization of an event.

Heitmann (2011, 49) addresses the dynamics of this concept, stating, "Staged authenticity implies that authenticity means the same to everyone, like a label that can be attached to an object, subject or experience. However, authenticity is not a given, measurable quality, applicable to a particular event or product, nor is it a fixed, static concept; it is negotiable, depending on the individual tourist and his/her perception of authenticity." Concerns about the commodification of culture, or the commercialization of cultural heritage, are a prime example of concerns about authenticity and community celebrations (Antony, 2014; Mbaiwa, 2011). Events must reach out to all segments if they are to be authentically representative of the community (Tisdall, 2013) and benefit all segments, too. Tourism benefits are rarely spread evenly across all community segments (Cole & Morgan, 2010).

Community Celebratory Events: A Thousand Points of Light

Mair and Whitford (2013) find that the literature on event definitions and types has received increased attention, reflecting the role they play in shaping discussion on the subject. However, socioeconomic impacts of community events, and the relationship between events and public policy agendas, have not benefited from closer attention. O'Sullivan, Pickernell, and Senyard (2009) provide a counterargument to those advocating for an even greater emphasis on improving process and measuring economic outputs rather than on socioculturally focused evaluation of celebratory events. This necessitates greater scholarship to maximize how community events emerge and evolve.

Art and celebratory events can be vehicles for shaping community identity because these are "concrete" ways of bringing people together in a common task (Derrett, 2003, 2004; Montero, 2015; Tucker-Raymond, Rosario-Ramos, & Rosario, 2011). Community development is impossible without developing and affirming a community's socially constructed sense of identity, based on respect and understanding of commonalities (Ghasemzadeh, 2013; Gunsoy & Hannam, 2013; Patterson et al., 2011; Ziakas & Costa, 2010a).

Community celebratory events can be considered critical vehicles within the arts creative industry, with potential for outcomes spanning all forms of community capital (Linko & Silvanto, 2011; Moeran & Pedersen, 2011). Chang and Mahadevan (2014) advance the interconnectedness and intrinsic value of business and the social benefits of arts and culture, making events a community fixture rather than a fad. Developing innovative ways of connecting with other practice fields increases the likelihood that events can gain prominence and funding, drawing the attention of scholars and practitioners not normally associated with events.

Community celebratory events share many similarities with artistic and cultural activities, and even sporting events, providing researchers and practitioners with a window for understanding and reaffirming community context and the importance of place and space in life (American Planning Association, 2011; Guazon, 2013; Otero, 2010). Sjollema and Hanley (2014) describe the use of poetry as a community development tool in community-based writing groups in Montreal, Canada, and its place within a constellation of the arts and celebratory events. Poetry contests can be part of a celebration, increasing participation for segments of the community that normally would not have an outlet for this form of art and creativity. Talent contests with various age categories, too, can be an added dimension to celebratory events, adding participants.

Celebratory events can be viewed from a collective learning perspective, adding a new dimension to the subject of crowds and celebrations. Roy (2013) comments on the role and importance of collective learning, comparing cultural events to documentary film festivals, to show how event organizers serve as catalysts in varied roles that can foster media literacy and critical thinking. Collective learning differs from individual learning because of the role and importance of groups sharing and shaping the moment of this accomplishment, and how the group validates and reinforces the learning, further increasing a community's identity and collective memory.

It would be a mistake to narrow a definition to just arts or culture since celebratory events encompass other critical dimensions and goals for a community's well-being, serving to engender civic participation across all segments of a community and enhancing the competencies (capital) of those involved in staging an event (Percy-Smith & Carney, 2011). A definition expansion brings challenges in understanding, researching, and evaluating events – but also brings its share of rewards.

Celebrations can occur for a range of reasons, with policymakers and other governmental entities often playing a facilitative, and at times dictatorial, role in planning events. These celebrations will consist of multiple developmental stages (Linko & Silvanto, 2011). Typically, the three distinct stages of an event are preparation, the actual event, and the creation of memory (Nuere & Ortuzar, 2013). The latter stage often goes unrecognized in the literature, but it is so important in communities where positive collective memories are rare. Each stage provides scholars with a window into these events, with distinct goals, activities, artifacts, rewards, and challenges. The whole must not be lost through a focus on the individual parts. Yet an understanding and appreciation of the individual parts is essential and worthy of attention.

Promoting the achievements of local institutions, such as libraries or small community museums, are common themes in community celebratory events (Silverman, 2010; Skot-Hansen, Rasmussen, & Jochumsen, 2013; Wronska-Friend, 2012). Individuals who work in these institutions can assume influential roles in the different stages of these events (Arnoldi, 2006). For example, librarians can be part of efforts to stage events and can determine how the events can be recorded and how the records can be saved and distributed (Smallwood, 2010). Librarians, in addition, can help research customs, music, art symbols, and other cultural manifestations that can be a part of events, as well as using other institutional resources in service to celebratory events (Braquet & Westfall, 2011).

In describing celebrations accompanying the deaths of well-known people, such as in the case of Jamaica's bling/dancehall funerals, Hope (2010) explores a version of celebratory events that has not received much attention in the scholarly literature. Día de los Muertos (the Day of the Dead) is another example; it is an annual event in Latin America, although it has found its way into the United States, which brings families together at ceremonies to honor their dead loved ones in a culturally affirming manner (Marchi, 2009). Although this event is individually focused, families in proximity to each other at the grave sites interact with each other and bring elements of festivals (Dias, 2005). Social and cultural capital, in this instance, carry over since many of these families see each other on a yearly basis and become a part of a unique collective experience and memory with deep cultural roots. Connectedness and place are intertwined in a manner that can also be found in community celebratory events (Samadi, Yunus, & Omar, 2012).

The Festa del grillo (the festival of crickets) in Florence, Italy, is credited as being one of the world's oldest insect festivals and is an example of the range of possibilities for creating a festival and how local circumstances dictate how events unfold (Hvengaard et al., 2013). The same festivals in different locations can result in different manifestations and outcomes, illustrating the power of context.

There are festivals devoted to the arts; birding, which is an increasingly popular form of celebration (Lawton & Weaver, 2010); gardens (Benfield, 2013; Ryan & Wollan, 2013); shopping and fashion (Entwistle & Rocamora, 2011; Peter, Anandkumar, & Peter, 2013; Skov & Meier, 2011; Weller, 2013); and sweets, such as the Eurochocolate Festival (Perugia, Italy) that appeals to chocolate lovers (Chirieleison, Montrone, & Scrucca, 2013). Food-related festivals, are very popular, whether they focus on a particular type of food – such as mussels, as in the case of rural North Jutland, Denmark (Blichfeldt & Halkier, 2014) – or a cuisine or food group more generally. Wildlife festivals promote a range of social, educational, recreational, economic, and community development goals, including conservation efforts (Hvenegaard, 2011).

Festivals lend themselves to celebrating a variety of events that can generate multiple forms of capital beyond the community hosting the event, including development of lasting collaborative partnerships that can be tapped for future projects (A. Smith, 2010). India's Toilet Festivals address local water and sanitation concerns, particularly for those living in marginalized sections of Mumbai and Pune. These communities have created festivals related to social justice to engender social activism and generate publicity – arguing that defecating in private is a right (McGeough, 2013). Other celebrations may mark tragic events – "original memories" – endowing these events with profound social meaning and providing residents with a chance to honor and celebrate the lives of those lost to those tragedies (Revet, 2011). These events create social bonding capital that can transform strangers and neighbors into friends, and connect generations in a way that may not otherwise be possible (Powell et al., 2011). Community celebratory events, in similar fashion to community music, will find definitions that are overly broad or very specific to be of practical use for practitioners, or in the case of the former, to researchers (Higgins, 2012; Schippers & Bartleet, 2013).

The above examples of festivals demonstrate the need for flexible definitions when discussing community celebrations. Local circumstances can be expected to wield considerable influence defining what constitutes a celebratory event and the level of support by residents, complicating a process that is challenging to begin with. Furthermore, a domains approach to celebratory events can encourage research and facilitate practice through categorization of activities and decisions, while acknowledging that there is no one way of thinking about events. A domain can capture an aspect of a celebratory event that can be considered independent but still overlapping with other domains (Lashley & Morrison, 2013). Such an approach encourages practice and scholarship and facilitates an international and multidisciplinary understanding of how celebratory events share universal and particularistic characteristics, since it provides flexibility in how categories get operationalized to take into consideration local circumstances (Getz, Andersson, & Carlsen, 2010).

Fortunately, other concepts can be tapped in developing a nuanced and multifaceted understanding of celebratory events. Ashworth and Page (2011) focus their attention on cities and examine the concept of *urban tourism*, which encompasses more than celebratory events, and note that this concept has made tremendous conceptual and research progress, having viability for creating greater awareness of the multiple benefits and challenges for urban communities attempting to generate various forms of capital.

Much of the literature on community celebrations has a strong tourism lens resulting in a heavy emphasis on economic goals and outcomes such as hotel occupancy rates and money spent by visitors. A comprehensive and culturally nuanced perspective is much more significant than this narrow view, allowing for other ways of identifying and measuring benefits.

Although there is a growing body of literature on community celebratory events, there is still a gap that must be filled if events are to continue to grow in importance. The role of volunteers in celebratory events can benefit from scholarship particularly when civic engagement is so prominent, and a key element of what is referred to as *community capital* (R.Murray, 2014; Zahra & McGehee, 2013). Raj, Walters, and Rashid (2013) apply Maslow's hierarchy of human needs to determine why people volunteer for celebratory events. Preventing volunteer burnout is a concern that is quite prevalent but generally not addressed in the literature (Stone & Millan, 2011). Murgante and colleagues (2011) report on the use of volunteers to gather information and map tourism services, illustrating the kind of information that can be obtained on events through civic engagement. Residents can represent many different backgrounds, with each bringing a fresh perspective and playing prominent roles in shaping future event staging (Nakano, 2001).

How are volunteers (human capital) recruited and supported to maximize their talents? What specific talents or qualities are sought and how are they sought? Does engagement (social capital) differ between different types of celebratory events? What role does the social network play in fostering civic engagement in these events? What are the perceived impacts of events on the organizers of these celebrations (Gursoy, Kim, & Uysal, 2004)? How have their competencies increased, and how have these gains been translated into other arenas benefiting the community? Answers to these and other questions facilitate increasing community participation and assessment of their contributions to celebratory events.

Transnationalism and Celebratory Events

There has been a dramatic expansion of multidisciplinary literature and research on the transnational experience (Schiller, 2012a; Zhou & Lee, 2013), bringing another popular concept to our understanding of community celebratory events. The role of cultural heritage and cultural events in the construction of a newcomer community and identity provides a window through which we can develop a more nuanced understanding of collective memory and how it facilitates or hinders public memory in the new homeland (Cardoso, 2013; Fricke, 2013; Ganeva-Raycheva, 2013; Hill, 2014). Zerubavel's (2012) book entitled *Time Maps: Collective Memory and the Social Shape of the Past* discusses how collective memory shapes how we remember the past, including the power of celebratory events in anchoring our sense of and place in the past. An ability to act collectively in new homelands is a hallmark of resilient transnational communities (Kotin, Dyrness, & Irazábal, 2011; Sarmiento & Beard, 2013).

There are no national or community boundaries that have kept community celebratory events out of their borders (Zuev & Virchow, 2014), and this bodes well for these events and the public rituals attached to them, especially for newcomers to a country (Francesco, 2014; Riaño-Alcalá & Goldring, 2014). Event traditions can connect old and new environments in a manner that integrates different age groups and facilitates the sharing of stories.

Cities (space and place) historically have provided refuge for immigrants, refugees, and individuals who are "different." In the case of newcomers who were immigrants or refugees this status brings ethno-national sentiments and possible conflicts from the original homeland, as they are transplanted in their new homes. Bakshi (2014) examines how new spatial settings harness these memories and translates them into narratives that reconstruct the past in the new settings, and addresses memory from various perspectives, including how memory is integrated into the basic fabric of a community and city. Bakshi's conclusions can be manifested through different celebratory events, which provide a vehicle for expressing historical narratives in a new geographical context. These authentic moments can be captured during community events and have profound social meanings for those staging and attending these events (Price, 2010), and if captured, they can be archived for future generations.

The importance of neighborhood historical projects and celebratory events that qualify as historically significant are increased in marginalized neighborhoods because of how their histories get recorded, often misinterpreted, and by whom, which is often outside of their purview and influence (Mooney-Melvin, 2014). Capturing history by recording (visual and/or written) celebratory events represents a way for communities to shape their own histories and craft their own narratives.

Miller (2014, 100) makes an important observation about acculturation and how the failure of newcomers to adjust to new surroundings can have negative consequences: "What creates the condition for tragedy is not disruption, which is generally seen as the negative aspect of migration, but continuity. In particular, the problems that arise from a desire to remain true to one's roots. The reason being that the customs, traditions and expectations that make sense and serve a person well in the context from which they come, may be singularly inappropriate, and betray that individual in the place they have come to." Miller's caution can be readily applied to celebratory events in newcomer communities. Events seeking to honor and maintain cultural traditions face challenges from residents who have difficulty relating to these traditions and stories of life in the home or "old" country. Tensions can result in disruptions in planning and supporting these events, further accentuating differences within the groups sponsoring these events and spilling over into other aspects of community life.

Histories are shaped by the sociodemographics of the individuals experiencing and recording them because they are viewed through their eyes and biases. Thus, children and youth view their presence within neighborhoods from a different vantage point than parents and older adults. Furthermore, multiple stigmas have a cumulative impact on how people perceive and socially navigate their surroundings. Individuals with high numbers of "undesirable" characteristics face the greatest challenges in having their needs met in an affirming manner.

Celebratory events bring immigrant communities opportunities for examining community social, political, economic, and cultural outcomes (Shutika, 2008). Zhou and Lee (2013) draw attention to the interrelationship between immigrants and organizations that support or represent them in the United States, and in the case of Chinese immigrants, how they strengthen the infrastructure and symbolic systems of ethnic communities and a community's capacity to generate capital related to the integration of immigrant newcomers. These organizations play critical roles in sponsoring and shaping how celebratory events evolve and are sustained over the years.

Community celebrations serve as cultural vehicles that facilitate transnationalism and the integration of newcomers into a host country (Bueltmann, 2012; Sandoval & Maldonado, 2012). Newcomer hometown associations (made up of immigrants originally from the same hometown) are often in a propitious position to sponsor celebratory events and local community development (Bada, 2014; Strunk, 2014). Salzbrunn (2014) examined Cologne, Germany's "No Fool Is Illegal" festival and traced its evolution over the centuries, as it incorporated new groups representing an increasing diversity within that city. Jukova's (2014) study of the Bulgarian Society of Western Canada found that the association is a valuable source of social and cultural capital for this community and reviewed the changes it has experienced since the initial settlement.

Solís, Fernández, and Alcalá (2013) address an unconventional viewpoint by focusing on the children of Mexican newcomers in New York City. Through community organizations and events, newcomer children are provided with an opportunity to experience the benefits of helping their community. These children and youth are considered community assets and community citizens. Adams (2013) focuses on how newcomer children develop a sense of place, their own connections to place, and the meaningfulness of these places, including what can be done to enhance these connections, which are excellent ways to engage newcomer youth. A transnational perspective brings a new way of viewing celebratory events, taking into account how global migration trends influence urban communities and local politics, with consequences for urban sectors with histories of welcoming newcomers (Erol, 2012; Ferdinand & Williams, 2013; McEachie, 2013; Waldinger, 2013). Newcomer communities turn to celebratory events for social, cultural, economic, and political reasons, as the events can serve as a bridge between the "old country" and the newly adopted country. Lewis (2010) reports on refugees and celebrations and how the events present participants with elements of homely familiarity and "strange novelty," but raise the need to view "community moments" from a nuanced understanding of the forces operating that led to their development.

There is an increasing body of literature illustrating the growing importance of community events, and the role they can play in creating an urban collective culture and civic affirmation in urban life and the neighborhood fabric of support (Amin, 2008; Betancur, 2011; Lee, Arcodia, & Lee, 2012). Celebratory events provide opportunities for collective participation, activism, learning, and experiences that are significantly different from everyday experiences (Huang, Li, & Cai, 2010), influencing actions and the development of collective memories that can be tapped in future endeavors. Perceived value, place attachment, loyalty, sponsoring organization, and visitor engagement, increase satisfaction with event attendance for those versed in community engagement theory and practice (Tsai et al., 2011; Kilanc, 2013; Yoon, Lee, & Lee, 2010). Jepson, Clarke, and Ragsdell (2014) argue that there is limited understanding, agreement, and research on engagement and participation in community festivals and events. Thus, much work in this area is needed if a capacity-enhancement paradigm will take hold with celebratory events.

Loyalty, and the creation of meaningful relationships between attendees rather than between attendees and the actual celebratory event, is critical in event staging, as discussed by Collin-Lachaud and Kjeldgaard (2013) in regards to French music festivals. *Loyalty* can be a very complex concept because of its multiple dimensions (Deng & Pierskalla, 2011; Wan & Chan, 2013). Lee (2014) argues that there is a paucity of empirical work on festival loyalty and that "atmospherics" have a positive, but indirect, impact on loyalty through positive emotions, satisfaction, and psychological commitment. Addressing the issue of loyalty is crucial to capturing the significance of an event and evaluating its impact on attendees.

Communities are ever changing and they are influenced by local and international forces and circumstances. Shifting demographics are an integral part of urban communities and can be considered dynamic and positive rather than a source of alarm, although shifting demographics may cause political upheavals in newcomers' countries of origin. Newcomer communities have rarely known a static existence as to who lives within their borders and how long they stay. It is important to look at those who move away, however, and consider how far they move and how frequently they return to their original community.

Urban communities undergoing demographic transformation due to the influx of newcomers, or as the result of gentrification, create places and spaces or "urban enclaves" (Stavrides, 2013) that either reinforce historical community identities or resist the imposition of new identities that might shift the power dynamics (Barthel et al., 2014; Bell, 2014; Bonilla, 2012; Main & Sandoval, 2015; Patel, 2014; Scambary, 2013). Groups contest these gentrified urban spaces, causing tensions and possible upheavals (Donish, 2013). Even when the nature or composition of a community changes, memorials can remain to celebrate the lives of individuals who would otherwise be forgotten due to time and the introduction of new population groups (Langegger, 2013). As a community changes due to demographics and other social forces, it becomes important that it take stock of where it has been, where it is, and where it hopes to go in the future. A community in transition can tap and mobilize assets, including maintaining traditions, to ground its identity within a shifting population base (Colding & Barthel, 2013), with celebrations playing an important role in accommodating newcomers and giving them voice.

Tracing the origins of a celebratory event can help a community develop a historical understanding and appreciation of a present-day event (Wakimoto, Bruce, & Partridge, 2013). Event archival information helps communities ground their experiences historically (Ferdinand & Williams, 2013; Smajda & Gerteis, 2012), and understanding historical antecedents helps present and future attendees appreciate the significance of a community event.

This contextualizing of the transformative journey is critical in helping a community develop its unique narrative, and helping the external community appreciate that story. Communities shifting in composition as long-time residents move out and are replaced by newcomers of different ethnic or racial backgrounds will experience tensions and conflicts. Reflecting on these changes will allow the community to define itself and to appreciate the process it went through to get to that critical point.

Festivals can assist communities to develop narratives that are nostalgic, further enhancing the experience of attending a celebratory event (Holyfield et al., 2013). Methods that can be used to reflect on this journey are much in demand, and events are important mechanisms in helping communities to accomplish this goal (Ecoma & Ecoma, 2013). These events also bring a dimension that can enhance the competencies of nonresidents, service providers, and educators who attend.

School teachers can participate in festivities and learn about the cultural traditions of their students and their families, including the values of inclusion and relevance, which can then be incorporated into the curriculum (Ginsburg & Craig, 2010). Grounding curriculum within students' cultural and lived experiences in this way makes content more interesting and meaningful. Human service providers can develop social networks that can be mobilized or incorporated into programs and services including outreach strategies tailored for community segments; this is well illustrated in fairs (Speiser, 2014).

Crowds and Community Festivities

Word associations with community celebratory events will uncover many different responses, including the word *crowds*. Celebratory events equate almost automatically with crowds. Crowds are associated with excitement and energy, as well as unpredictability and even danger (Stohl, 2014). Kendrick and Haslam (2010) argue that crowds are an integral part of our daily lives, but there is limited understanding of how these experiences can be enhanced when it comes to community celebratory events.

Snow and Owens (2013) discuss the ubiquitous nature of crowds and their importance in shaping historical events, from the storming of the Bastille in 1789 to the sit-ins and marches of the civil rights movement in the twentieth century. Crowds can be manifested in other forms with social and cultural implications that go beyond politics.

Crowds and celebrations are closely tied together, and they are popularly thought of in terms of numbers, although estimating crowd size is extremely challenging (Vaccari et al., 2010; Watson & Yip, 2011). Crowd size is a key indicator of the success or failure of an event. It seems like a news story of an event is not complete without a crowd estimate and a comparison with the previous year's numbers. Crowd characteristics, however, are often not reported nor are the characteristics of those who do not attend, which can be considerably different (Milner, Jago, & Deery, 2004). However, attendance does not necessary mean "actual" or "physical" attendance. Showing events on large digital screens increases the reach to those who cannot attend an event (Koeman et al., 2014). Efforts to judge the reach of an event must take into account these types of participation as well.

Although the decision-making process of attending events is often viewed from an individual perspective, this is not how it is conceptualized in this book because events are closely associated with crowds. Crowds and groups are not mutually exclusive of each other, and can be complimentary, adding a nuanced understanding. When crowds are tied to urban settings and events, crowd management takes on added urgency (Al-Kodmany, 2013). Crowd simulations have increased in importance as a way of reducing the chances of injuries and deaths from mass events (Klein, Köster, & Meister, 2010).

This feature of participation has not received sufficient attention in the professional literature, and when addressed, it invariably focuses on negative health and safety consequences (Batty, Desyllas, & Duxbury, 2003b; Hutton et al., 2012; Hutton, Brown, & Verdonk, 2013; Ma et al., 2013), or relates to dangers associated with public demonstrations (Fillieule, 2012; Wintemute et al., 2011). Identifying potentially dangerous situations is an important goal for event organizers (Feng, 2013; Wirz et al., 2012). How communication involving mobile devices can be disrupted in large gatherings has received attention because of potential safety concerns and getting emergency vehicles to the scene (Barzan et al., 2013).

Aggressive behavior is more prevalent in drinking crowds when compared with nondrinking crowds (Moore et al., 2008). Aggression and drinking alcohol take on greater significance in community celebratory events. Large, potentially aggressive crowds may deter those afraid of crowds from attending an event because of the potential for harm and unpleasant experiences (Boo, Carruthers, & Busser, 2014). Events with reputations for drinking, acting-out behaviors, and arrests will keep away families with young children or older adults, and cater to a narrow age range. The selling of alcohol at an event can exasperate alcohol-related violence and behaviors.

Crowds can be viewed from a multifaceted perspective. Physical, environmental, and psychological perspectives stand out because of how they tap distinct viewpoints that lend themselves to multidisciplinary attention (Hutton et al., 2012), and they can paint a portrait of how their presence influences events and how events shape crowd reactions. This interrelationship is dynamic and arduous to capture using conventional research methods, but if captured, exciting for the field of celebratory events.

Concerns and Paradoxes of Community Celebratory Events

Community-sponsored celebratory events provide an opportunity, place, and space for residents to take stock of and generate multiple forms of assets or capital. However, host communities can experience stress and other consequences when hosting events (E.J. Jordan, 2015). Some public spectacles have been criticized for being highly scripted, prescribing public spaces and their meanings, and being used for political and economic gains for a select group (Stevens & Shin, 2012). For example, some marketers use events to try to sell their products, and this introduces strong economic or commercial motives that can undermine or overwhelm other forms of capital. The marketing could even compromise the long-term health and well-being of a community: The New York State Department of Health's report entitled *Exposure to Pro-Tobacco Marketing and Promotions among New Yorkers*, prepared by Loomis, Nguyen, and Kim (2011), illustrates how certain industries, which have been labeled "merchants of death," see community events as ideal venues to advertise for their products. These events draw large crowds so marketers can target a pre-selected group (a very efficient and effective use of marketing resources), often using cultural symbols to make their products even more attractive.

Cinco de Mayo celebrations marking the independence of Mexico from France represent an excellent example of how an event has been taken over by commercial or corporate interests (Hayes-Bautista, 2012, 189): "In the second half of the twentieth century, big business also discovered the Cinco de Mayo's summoning power. Companies saw the holiday's public celebration offered an excellent opportunity to expand into the Latino market via the sponsorship of musical or other cultural events. Consequently, by the 1980s, corporate influence was noticeable in the holiday celebrations." Events have an evolutionary history that generates different capital/assets over an extended period of time, and also generates unhealthy consequences (Goh, 2011, 2013).

Events that are associated with alcohol consumption can be unwelcoming for children and families (Castro et al., 2014; Németh et al., 2011). Celebratory events can become a magnet for public intoxication and unruly, even life-threatening anti-social behavior (Deery & Jago, 2010; Hollows et al., 2013; Pennay & Room, 2012). The use of alcohol and illicit drugs can cause safety concerns (Barzan et al., 2013). Alcohol consumption can lead to risky sexual behavior at these events (Choudhry et al., 2014; Tan et al., 2014; Tumwesigye et al., 2012), including sexual assaults or the spread of sexually communicable diseases, which can tarnish of the reputation of an event. Ikuomola, Okunola, and Akindutire (2014) provide a vivid description of potential safety and health concerns pertaining to carnivals in Lagos, Nigeria, for example.

Bagri (2014, 7) reports on the Hindu festival of Dahl Handi in Mumbai, India, in which a human pyramid of boys aged 4 to 10 years marks the homage to the god Krishna, and suggests a relationship between crowds, safety, and the role of corporate sponsorship: "The *Indian Express*, a daily newspaper, reported Tuesday that 202 participants were hurt in Mumbai's celebrations. In addition, a 14-year-old died earlier this month after failing from the top of a five-tier pyramid during a practice. In recent years political parties and corporations have jumped in as sponsors, and the pyramids have reached improbable heights, with the current record being a nine-tier pyramid, just over 40 feet high. Prize money for the tallest pyramids can go up to 10 million rupees, or $164,000."

Other risks at community celebratory events can be food-borne illnesses (Gaulin, Lê, & Kosatsky, 2010; Wilson, 2013) or exposure to second-hand smoke or toxins that may be part of beads and other paraphernalia used in the festivities (Mage et al., 2010; Mielke, Gonzales, & Powell, 2012). Smoking at outdoor events compromises air quality by exposing attendees to fine particulate matter, as revealed in a case in Edmonton, Alberta (Collins, Parsons, & Zinyemba, 2014). Attendees with compromised health are more vulnerable to negative consequences from attending such events, and thus they may stay away even when the event may be of great interest to them.

Noise pollution is closely associated with many celebrations (Jamir, Nongkynrih, & Gupta, 2014). Part of the experience of attending an event is exposure to loud noises, and this often plays an influential role in shaping ambiance in large-crowd events. However, excessive noise at events in population-dense places can make it difficult for attendees to enjoy the moment (Poling & Thalheimer, 2011). Excessive noise can reduce attendance by individuals with sensitivity to noise, including children. When excessive noise is combined with other factors, a celebratory moment can become an unpleasant memory.

Some celebrations involve lighting bonfires, which can become a potential source of illness due to contaminants from the metal-containing materials used (Dao, Morrison, & Zhang, 2012), and they also can result in burns. India's Hindu festivals, such as Durga Pooja, Ganesh Chaturthi, Deepawali, and Holi, which are some of the most popular in that country, have been found to create environmental hazards due to the use, as an essential part of the festivities, of substances that are not biodegradable (Khan, 2013), with health-compromising implications for attendees.

Fireworks have a long tradition associated with festivals (Recio Mir & Cinelli, 2013), including at 4th of July celebrations in the United States. Fire can be a part of a celebration in other forms, for example, at the Burning Man Festival in the western United States (Bowditch, 2013), or during mass gatherings to celebrate an important sports victory (Hawkins & Brice, 2010). However, contaminants released by fireworks can have a severe impact on the quality of the ambient air and can result in airborne illnesses, as discussed in case studies of the traditional Vishu festival in Kerala, India, and the Chinese New Year celebration in Jinan, China (Nishanth et al., 2012; Yang et al., 2014). In addition, mild-to-severe ocular surface injuries caused by sprayed foam are not uncommon (Abulafia et al., 2013).

Soomaroo and Murray (2012) review the literature on mass gathering events from 1971 to 2011, and they conclude that these events have a tremendous potential to strain local health systems because of the size of crowds, limited physical access to event sites, crowd control, and limited on-site health care. Injuries and fatalities resulting from stampedes at festivals are a major hazard in some countries (Aron, 2011). For example, in India religious and pilgrimage events are responsible for 79 percent of all stampedes (Illiyas et al., 2013). Event organizers must consider the causes of unexpected mass crowd movements, particularly at events with a religious origin because these can attract groups that are against the sponsoring religious groups. Events in Northern Ireland are an excellent example of how one religious group comes out against another religious group and not just symbolically.

Religious celebrations can address stereotypes or reinforce them. In the latter case, a valuable opportunity gets wasted. Depending upon the goals of the organizers, they can actively seek to address stereotypes or reinforce them. If the goals are to educate outsiders, the events will transpire with this in mind. If the goals are internally focused, they will not make an effort to enlighten outsiders. In essence, we will use symbols that are important to us.

Community celebratory events can find their way into urban legend folklore. Detroit's Devil's Night (Chafets, 2013; McDonald, 2013) has received national coverage in the United States, if not internationally, because of the hundreds of fires that occur during the days leading up to and after Halloween. The destruction of private property combines with potential injuries to enhance the destructive force of events of these types. Botelho-Nevers and Gautret's (2012) description of the health consequences of attendance at events brings this point home, noting that open-air festivals in particular can increase the likelihood of transmission of communicable diseases such as "*Cryptosporium parvum, Campylobacter spp., Escherichia coli, Salmonella enterica, Shigella sonnei, Staphylococcus aureus*, hepatitis A virus, influenza virus, measles virus, mumps virus and norovirus." They also note that "sexual transmission of infectious diseases may also occur and is likely to be underestimated and underreported."

Large crowds can be perfect venues for spreading highly infectious diseases due to close contact (Mykletun, 2011). Event medicine has emerged as a field of practice in response to health consequences of attending mass gatherings (Chang et al., 2010; Livingston et al., 2014; Moore et al., 2011; Pepe & Nichols, 2013). The impact of various conditions at events on children has started to receive attention because of the unique challenges for this age group (McQueen, 2010).

Another concern is physical safety and psychological acceptance, which can be compromised at large-scale events; the potential of sexual assaults is one example of a risk to personal safety (Koopman, 2013). Markwell and Tomsen (2010, 225) researched the perceived and experienced risk, safety, and hostility by gays and lesbians at large-scale gay and lesbian festivals and special events in Australia, and found that LGBTQ attendees of the Sydney Gay and Lesbian Madri Gras feel comfortable and safe at most of the festival's events; however, these attendees report feeling unsafe or threatened at events that are larger in scale and have a wide heterosexual participation as well.

The association of suicides with community celebratory events has not been explored in the literature, although suicides and public holidays have been addressed (Barker, O'Gorman, & de Leo, 2014). Examining suicides immediately preceding and following a celebratory event brings a disturbing dimension that warrants further attention from scholars and practitioners alike. Stress-related illnesses, such as epilepsy, can be exacerbated during celebratory periods, and further study is warranted here as well (van Campen et al., 2014). A systematic examination of this phenomenon may open up a new perspective on celebratory events and how they can trigger post-traumatic stress disorders.

Events such as parades may reveal aspects of a community that purposefully seeks to engender hate, as in the case of the Ku Klux Klan in the United States and its counterparts in other countries. Their messages of racism, anti-Semitism, anti-LGBTQ, and anti-Catholicism as they seek to shape public opinion, recruit members, and solidify a movement convey to the broader community, and the country as a whole, a disturbing level of intolerance. Blee and McDowell (2013) call this the "duality of spectacle and secrecy." A community celebration can embrace feelings of hatred and distain. Thus, community celebrations are not always seeking to be inclusive and spreading a positive message.

Public perceptions and attitudes concerning LGBTQ communities have been studied. Pride parades can involve highly sexualized behavior, including fornication on floats (Waitt & Stapel, 2013). When this occurs, conflicting views emerge because of how stereotypes are reinforced. With regard to ethnic parades and festivals, commercialism and an emphasis on youthful looks have raised issues of ageism and celebratory events (Loukaitou-Sideris & Ehrenfeucht, 2009). Event organizers must be ever vigilant in considering how events reinforce or create new stereotypes (Browne, 2007). For example, what efforts are being made to ensure that residents with disabilities are included, making the events welcoming of all regardless of abilities?

This question generates discussions that can lead to all residents being a part of a community's celebration. Celebratory events provide communities with an opportunity to reach out to engage residents with disabilities (McPhedran, 2011), a group that can be found in all communities, and is often overlooked. These residents bring overlooked talents, so a comprehensive assessment of assets is essential to take into account all residents' assets.

A central goal is to have events become a prominent institutional part of a community's collective memory through its oral and recorded history. The institutionalization of events means that they have to achieve a level of sustainability and become part of a community's social fabric and memory. This goal does not mean that the event does not evolve and change with the times, or that it will be free of tensions and conflicts. Event institutionalization provides a community with a pre-established venue and opportunity to address important goals in the present and future.

As celebratory events achieve greater success they invariably expand in scope and complexity, necessitating the introduction of professional event planners (Gallagher & Pike, 2011; Silvers & Goldblatt, 2012). Professionals bring challenges for communities maintaining the original goals that led to the creation of an event without compromising their autonomy (S. Brown, 2014; Jiang & Schmader, 2014). Events evolve or simply die because of inertia and lack of local support. The former suggests the support necessary for sustainability; the latter may be inevitable and can be hastened through poor planning and inter- and intra-group conflicts and tensions. Events can be imposed on communities rather having them be organic, with deleterious social consequences representing another effort at cultural and place exploitation of communities that are susceptible to outside influences. This exploitation can involve active collaboration of major community stakeholders with external sources, giving the illusion that the celebration is community-centered.

The privatization and commercialization of public space undermines the autonomy of communities and their abilities to influence what events take place within their boundaries (K. Smith, 2013). Celebratory events bring increased surveillance/monitoring, serving to reassure attendees but also alarm those with histories of living under authoritarian regimes, for example (Boersma, 2013; Wirz et al., 2013). Newcomer communities with high concentrations of undocumented residents will be prone to avoid participating in events being monitored for fears of deportation. High publicity brings high authority vigilance, which can be highly undesirable by particular segments of the community, thereby excluding them, intentionally or unintentionally.

How to Kill a Community Celebratory Event

Celebratory community events can have negative impacts on sponsoring communities, from the costs involved or a broader negative consequence that might reduce the overall benefit of the event; it's important for event organizers to plan for and manage potential negative impacts

(Sharpley & Stone, 2011). Negative consequences can be manifested in a wide variety of events and areas. Festivals can cause traffic disruptions, significantly altering pedestrian traffic, create environmental pollution, and have countless other detrimental consequences (Bagiran & Kurgun, 2013).

It is fitting to end this chapter on a humorous note, although those who have experienced failure in staging a community event may not find this list to be humorous (Ward, 2008):

1. Rest assured that everything will fall into place. There's no need to organize.
2. Begin your planning tomorrow. These events are a piece of cake.
3. Ignore Health Department regulations. After all, the inspector was once on your bowling team.
4. Give everybody equal authority. There's no need for leadership.
5. Assume that publicity is under control. The local newspaper is sure to provide front-page coverage.
6. Draw up rigid plans. Flexibility is for gymnasts, not event organizers.
7. Forget the idea of a simple event. Get your money's worth and start out with a weeklong event.
8. Demand help from local businesses and organizations. They owe you some cooperation.
9. Don't worry about extra help. You and your six helpers can handle any crowd.
10. Move the event date around from year to year. There's no reason to establish a traditional time for it.
11. Discard receipts, invoices, and other records. These things just get in the way.
12. Let somebody else worry about start-up money. Spend your time auditioning the entertainment.

This list resonates with anyone who has staged an event and had it become a nightmare. The items on that list may appear as self-evident. Nevertheless, there is a good reason why they appear on this list (Carlsen et al., 2010).

The following chapters provide countermeasures that can help increase the likelihood of success, although no community event will unfold in a perfect manner; some disappointments are inevitable but can be minimized with careful planning (Pentecost, Spence, & Kale, 2011). Thus, the best that can be hoped for is to increase the likelihood that an event will unfold as planned. As the saying goes, to plan is human and to implement is divine.

Conclusion

Numerous terms and concepts have been covered in this chapter and may rightly raise readers' concerns about how community celebratory events ended up encompassing such a broad arena. The field is expanding, requiring the introduction of concepts not normally associated with this field. This expansion can be a concern or an opportunity, but if the latter, one with immense challenges and rewards.

This book draws from diverse practice fields and disciplines, synthesizing existing knowledge from academic and nonacademic sources. Newfound knowledge will originate from the community practitioners who have not found time for publishing or formally presenting their insights and recommendations at conferences. This book seeks to advance knowledge on a subject that has for too long been viewed from an economic/tourism perspective. Although that is an important perspective, urban communities are more complex and consist of more than economic capital; there are other forms of capital that must be tapped to achieve a comprehensive portrait.

3 Evaluating Community Celebratory Events

If celebratory events are to be conceptualized as community social interventions, evaluation must play a prominent role and it, too, must be conceptualized as an intervention (Getz, 2012). In addition, evaluation must be conceptualized in the same manner and importance as any other intervention, sharing similar values and principles. It should never be thought of as an added-on activity or phase because its importance is equal to any of the other activities in organizing an event (Foley et al., 2012; Schulenkorf, 2012). A chapter on this topic is thus warranted due to the importance of evaluation.

Evaluation will have an increasingly important role in the coming years as community celebratory events become a more active and meaningful part of community social work practice, as evidence-based practice takes hold in this field, and as external funds are sought from sources not usually involved in sponsorships, such as noncorporate sponsors. Evaluation takes on added significance when celebratory events are large and high publicity-intensive. Evaluation approaches, techniques, and methodologies will need to be modified to be community user-friendly.

This quest for innovation will be exciting and rewarding, but it will also result in tensions and challenges. This chapter, although not comprehensive, addresses aspects that demand specific attention, and evaluation will be addressed throughout the remainder of this book. Sullivan (2014) argues that the ideal event prototype is one that has a clear emotional effect on a group and is often one in which there is a positive outcome for most participants. Such positive outcomes, however, are influenced by a myriad of factors, and evaluation will help solve this puzzle.

This chapter examines the following seven interrelated topics that seek to identify the positive outcomes of a celebratory community event and how best to achieve them: (1) an overview of the state of celebratory evaluation and research; (2) rethinking evaluation and community celebratory events; (3) the process and focus of event evaluation; (4) areas for evaluating the impact of celebratory events; (5) challenges in conducting evaluation; (6) innovative strategies for evaluating community celebratory events; and (7) evaluation reports and their dissemination. Each section highlights the most salient points for generating discussion and possible solutions.

An Overview of Celebratory Event Evaluation and Research

The evaluation of community celebratory events is a relatively new topic, particularly when viewed from a historical context dating back to the beginning of recorded time (Fox et al., 2014). This is not to say that serious efforts at assessing factors associated with the success or failure of events have not been undertaken. However, there is no denying that the more we learn about this subject, the more complex it becomes (Jayaswal, 2010). Mair and Whitford (2013, 15) sum up the research to date on evaluating celebratory events:

> For too long, the evaluation of events has been limited by an emphasis on assessing the economic benefits despite the fact that the methodologies adopted in such research are increasingly questioned ... Since the 1980s, however, there has been increased interest in sustainability and consequently, increasing calls for a broader approach to evaluating events ... Certainly the impacts of events have continued to be the focus of much research attention, and the socio-cultural and environmental impacts of events appear to be gaining in importance. Indeed, socio-cultural and community impacts (including resident attitudes to events, social capital and social inclusion, community pride, etc.) were identified in this study as being the most important topic for future research, with both environmental impacts of events, and the sustainability of events also featuring in the top five.

The evaluation of celebratory events has been undertaken in the past, yet it has been limited in focus and scope, compromising our understanding of the impact of events on the communities sponsoring them. In addition, models guiding these evaluation efforts have stressed professionally led efforts, often with minimal or no involvement by the community – a critical element in any evaluation effort when using a capacity-enhancement paradigm. These evaluations have not moved the field forward in a manner to achieve its potential for transforming a community.

Rethinking Evaluation and Community Celebratory Events

Taking a capacity-enhancement approach that relies on community capital/assets necessitates a paradigm shift to capture the multifaceted impact of events, requiring evaluators to turn to methods that historically have not been used with community celebrations, and even developing new methods that are particularly sensitive to celebratory events, incorporating new technologies in the process. Capturing social and cultural nuances is important because of how local circumstances can spark use of events to meet social, cultural, political, and economic needs. Presenting a multifaceted and nuanced picture can awaken interest in this social intervention.

Taking stock of where the field of event evaluation has been and should go in the future is often left in the hands of academics because we have the time for undertaking this task, and we are paid for it, which is no small matter. Community practitioners and residents are not in such a propitious position to undertake this type of analysis. Moreover, there is no unified vision of how to evaluate celebratory events. Nevertheless, there are distinct and thorny issues that simply keep appearing in the literature.

The critical question of "So what?" emerges in any discussion about the promise of celebratory events, and it stands to reason why this section starts with this question, otherwise referred to as "impact." Evaluating the impact of celebratory events is compounded by controversies surrounding impact analysis, with the added dimension of inherent ambiguity in events making the project of evaluation that much more challenging (Mangia et al., 2011, 1020): "The main problem is represented not by simply how to measure but what to measure, thinking that we can develop the right tool, after defining the object of the evaluation process."

Determining what and how to measure in evaluating celebratory events, and the role of the community in this evaluation, is rooted in values, and these values may not be explicit in the literature, further compounding communication and direction finding for future evaluations. This section provides a broad understanding of celebratory event evaluation, including inherent tensions and controversies. Event evaluations share commonalities and rewards and challenges, including the need for innovative approaches and methods that do justice to these community events.

When there is an evaluation of impact, it tends to focus on economic impact because of an overemphasis on economic development; this is an important yet narrow view of celebratory events because of the increasing presence of celebratory events and a growing body of scholarly activity related to them (Small, Edwards, & Sheridan, 2005). Celebratory events are about more than economics, and it would be a disservice not to evaluate their social, cultural, and political appeal and impact (Andersson & Lundberg, 2013).

Broadening the parameters of research and evaluation of celebratory events to include the noneconomic impact brings both rewards and challenges for evaluators (Andersson & Lundberg, 2013; Rogers & Barron, 2010). A shift away from economic capital to include other forms of capital expands our understanding of how celebratory events transform urban and highly marginalized communities, beyond economic outcomes, while also serving as an impetus for using events as social interventions that rally communities by affirming their positives, or assets.

One of the primary challenges evaluators face is to ask questions that have not been asked in the past and to generate data where data never existed; this can be both time and labor intensive, but is necessary to develop a comprehensive and nuanced understanding of community celebratory events (Vargas-Hernandez, 2012). Evaluation can be thought of as storytelling, and all stories are culturally and contextually grounded, emphasizing experiential knowledge that can inform interventions (LaFrance, Nichols, & Kirkhart, 2012). Forging ahead into unchartered territory to garner these stories requires a willingness to entertain innovative approaches.

A paradigm shift to appreciating how celebratory events help communities is a critical starting point for enhancing the value of such events. Scerri and James (2010) suggest using indicators that reflect the community's perceptions and input to help ensure that they influence the sustainability of the events. Tapping the community's perceptions and expectations yields perspectives that may well be different from those of academics or practitioners. How these differences are obtained and addressed captures the meaning of events for the communities that stage them, determines how academics can report this to audiences in a position to foster and/or sponsor events, and helps prepare future community practitioners.

Posing fundamental questions that a community wants answered is an appropriate place to start when event evaluations are community-centered

(Langhout, 2014). These questions may not be the ones that funders and/or sponsors are interested in asking, and compromise may be needed to ensure that both the community and the external stakeholders get answers to their evaluation questions. Meeting community needs without losing sight of funders is a skill that community practitioners endeavor to acquire and enhance.

Although event organizers should have a specific demographic profile of their target group and goals, who actually attends an event is unpredictable, and this makes evaluation of the impact more arduous. If a particular demographic group was targeted and its members did not attend, evaluators need to ask them, as well as those who did attend: why? This unpredictability makes multifaceted evaluation essential. These and other challenges that require the development of innovative methods that can capture crucial nuances will be addressed in the following chapters.

Process and Focus of Evaluating Celebratory Events

Community social work practitioners may argue that practice is all about process. However, the perennial "bottom line" question cannot be ignored. Evaluation must focus on both process and outcome. Event evaluation needs to embrace a community-centered process and answer questions of relevance to the community, and with an appropriate degree of rigor involved in determining the outcomes (Davies, Coleman, & Ramchandani, 2013). These questions may be different from what community practitioners and funders may think is important.

Community involvement in the initial planning is the most appropriate time to start asking questions. Balancing community and funder needs will be a delicate undertaking, and although that will not be a new experience for seasoned evaluators, it may well be for communities without histories of external funding for their events. Expert guidance can enhance the capacity of communities for planning future events. No facets of events must be overlooked in evaluating their impact and success. Evaluation must also address how events were marketed and to whom (Wood, 2009), creating a very detailed understanding of who should attend an event and why (Yoo, Lee, & Lee, 2015), including the wording used in marketing. However, the fundamental questions are, who are the intended attendees of this event, and what is it the organizers want them to leave with?

Chances are good that there is no one group or profile of who the intended attendees of an event are (or were), making marketing and evaluation that much more arduous. In all likelihood there are primary and secondary groups that are expected to attend. Gender, however, is usually overlooked as a factor impacting attendance. Lahiri-Dutt and Ahmad (2012) point out that in the efforts to understand the impact of an event, gender is often not addressed. Gender takes on even greater significance in cases where ethnic or racial groups have long traditions of excluding women from decision-making roles.

Each of the six factors addressed in this section focuses on a particular aspect of evaluation that aids community practitioners and academics alike in thinking how event evaluation can be undertaken. These factors are not unique to events; however, in isolation and combination with each other, they present unique challenges in evaluating community celebratory events.

Community Participation and Ownership

The increased prominence of communities in sponsoring and running celebratory events necessitates that communities, too, play a prominent role in their evaluation. Events that are not community-centered with the active participation of residents cannot result in community ownership (Reid, 2004). Further, ownership is not age-restricted. A school celebratory event should have students play an active role so that it is truly student-centered (Kirk-Downey & Perry, 2006). The concept of ownership should not be restricted by other demographic factors.

A paradigm shift in epistemology, from one that is educational-expertise centered (i.e., on those with university degrees) toward one emphasizing the community's residents and participatory principles (i.e., consumers, or experiential expertise) increases the likelihood that social change will result from a celebratory event (Hutchison & Lord, 2012). When marginalized communities stage their own celebratory events, multiple goals must be articulated that can help transform these communities. Building in a high degree of research and evaluation based on community participation in staging the event makes this result more likely (Ross et al., 2010).

Cultural and Social Capital as Core Elements

The historical emphasis on evaluating the economic benefits of community events, because of an overemphasis on generating tourism-related money, has shortchanged other forms of capital, and the professional literature bears this bias out in the sheer volume of published items devoted to the economic benefits of events. However, in fact, cultural capital and social capital stand out in importance, particularly when celebratory events are focused on newcomer, ethnic, and racial communities and their struggles to achieve social justice.

Although the prominence of social and cultural forms of capital is well acknowledged and often closely associated, even to the point where they cannot be disentangled, evaluating their impact on a community event remains challenging. Determining how these forms of capital influence other types of assets becomes significant in evaluating celebratory events. One could argue that an event is not possible without cultural and social capital, particularly when taking into consideration the synergistic impact of these two forms of capital.

Community-centered events have not typically been conceptualized as fundraisers, although money can be an attractive benefit, unless the event is driven by financial sponsors. Fairs cannot be sustained without social capital playing a critical, if not central role – making social capital more significant than money in the sustainability of fairs (Ball & Wanitshka, 2014).

Qualitative/Quantitative Dimensions and Outcomes

The complexity of community celebratory events is such that no one method of evaluation can do justice to all of the elements and dimensions involved in staging and ensuring their ultimate success. This is particularly the case when one of the goals of an event is to be welcoming of all segments of a community.

Comprehensive evaluation is difficult and expensive. Qualitative methods facilitate the gathering of highly personal narratives on event experiences (Price, 2010), or what is referred to as the *ambiance or nuanced approach*. Qualitative approaches can use various methods, some of which are conventional and others unconventional and emerging in the field.

Devine, Moss, and Walmsley (2014, 99) highlight the intangibles in determining the outcome of a celebratory event: "Events are typically made up of mostly intangible offerings, which may include (but are not limited to): atmosphere, audience interaction, decoration, theming, entertainment, venue aesthetics and various additional novelties. Each individual attendee will experience the event in a different manner depending on their personal perceptions and expectations." These intangibles cannot be measured using one method or technique.

Langen and Garcia's literature review (2009) on measuring the impact of cultural events suggests that qualitative methods (including participatory mapping) are the most appropriate methods for measurement, but they also highlight the importance of combining qualitative with quantitative methods, or a mixed-method approach. Their review mentions that most evaluations of the impact of major cultural festivals are geared toward organizers and funders, and the common approach is analysis of "visitor expenditure data."

The lack of evaluation data that systematically addresses the negative monetary consequences of festivals raises important questions about how these negative ramifications can be overlooked. This is important since economic goals and outcomes can be a central goal for initiating a celebratory event, keeping in mind the goal of tourism (money) and "branding" as key motivators, particularly when led by major political and economic stakeholders.

Resources, Financial and Nonfinancial

Resources play an important role when undertaking the evaluation of a community celebratory event. Some event planners would argue that a lack of resources is often the primary reason why events are unsuccessful. However, it is important to conceptualize resources as both financial and nonfinancial. Financial resources will dictate the emphasis and level of the evaluation, allowing for resources to tap local commercial establishments in the hopes of sharing funds within the community.

Nonfinancial resources can be conceptualized in a broad manner with a key goal of obtaining buy-in and support of community events. These resources can involve using local artistic talent in publicizing evaluation efforts through the development of leaflets, posters, and digital images, for example. Local establishments can serve as meeting places. Involving community volunteers to become a part of the evaluation is an excellent way of involving the community and enhancing human capital.

Civic Engagement and Volunteering

As already noted, resources used in celebratory events can be both financial and nonfinancial. Tapping civic spirit can be considered a nonfinancial resource that in many ways is far more important in the life of a community celebratory event than financial resources. Engaging the local populace brings a dynamic element to an event that is community-centered, emphasizing the values and principles of participatory democracy. Obtaining community involvement is easier when the participation is focused on a positive project that will elicit pride.

Nonfinancial resources can play a role in celebratory events by facilitating contributions from a variety of sources. A heavy emphasis on civic engagement as an event theme increases the likelihood of an event being community centered and community owned. Getting volunteers must be systematically planned because it takes considerable resources to run a successful volunteer outreach effort at these events. Volunteers must be recruited, screened, trained, supervised, and validated, bearing in mind that they represent an investment in the future.

Involvement of All Significant Segments of the Community

The engagement of all significant segments of the community is a worthy social goal. No community has the "luxury" of excluding an important segment of its population from participating in the life and celebration of that community. Participation is a central goal in democratic societies. Civic engagement takes on even greater significance when discussing a community with numerous social and economic challenges that make survival itself a goal, and where there is a history of disenfranchisement. Excluded communities must not experience further exclusion within their own boundaries (Delgado & Humm-Delgado, 2013).

All segments of the community can contribute some form of capital. Some segments have financial resources, while others may have insights or a grasp of cultural traditions that more acculturated segments no longer have. Involving all significant groups helps ensure that celebratory events have the support of the entire community, increasing the chances of sustainability because no one segment "owns" the events.

Sustainability necessitates that communities be creative in staging events. Parades and festivals can be sponsored by two or more ethnic or racial groups although this is highly unusual; for example, in Somerville, Massachusetts, the Haitian and Brazilian communities joined together for a parade and festival in 2014 (J. Smith, 2014). These communities share an African heritage and address common goals and concerns; Haiti and Brazil formed a special relationship when Brazil was the first nation to donate help in Haiti after the 2010 earthquake there. Describing the primary purpose in this case, J. Smith (2014, B2) writes: "About a month in the making, the festival was 'not just about entertaining people ... It's important for the immigrant communities to know that they are welcome and safe.'" Sustainability increases through cooperation between community segments.

Areas for Evaluating the Impact of a Celebratory Event

An event evaluation must gather data on obvious and nuanced outcomes, and do so within a local-cultural context, because it is impossible to have a comprehensive understanding of a community event without this grounding. An event evaluation must answer seven critical questions that will not be unfamiliar to community practitioners: (1) why? (2) where? (3) when? (4) what? (5) how? (6) how many? and (7) who? Brown and James (2004) and Goldblatt (1997) argue that these questions are not just technical but also tap core values. To say that evaluation is challenging would be an understatement because of how local events can appeal to a wide cross-section of the community, particularly in urban and multiethnic and/or multi-racial communities.

Fredline, Deery, and Jago (2013) emphasize the importance of understanding the long-term impact of an event on residents' quality of life and raise the importance of longitudinal studies that measure opinion changes over an extended period of time rather than focusing on immediate benefits.

The following factors are illustrative, although not exhaustive, of the types of information that can be considered important in developing an informed understanding of how an event evolves and impacts the host community:

Economic. Consider the opportunities for tourism, commercial outcomes, employment, inflated prices, interruption of normal business, housing disruption, and increase in local entrepreneurial culture.

Environment/Physical. Be aware of any damage, noise pollution, overcrowded foot traffic, transportation, enhancement of quality of life, urban renewal, restoration of public landmarks, infrastructure, buildings, standard of public facilities, and the public awareness raised by environmental organizations.

Political. An event can be a platform for industry and economic development, a catalyst for political reform, whether for the event holder or the stakeholder, and an influence in driving certain issues or ideology.

Social. Consider whether community cohesion increases through the establishment and enhancement of social relations as a result of attending and participating in the event, but also that the disruption of relationships can occur, too.

Sociocultural. There may be disruption to residents' lifestyles, which has to do with festival management issues, which can lead to vandalism, crime, other forms of acting out behavior, and/or overcrowding, but also consider group representation, festival/community cultural identity, increased awareness, and educational awareness.

Community identity. Positive community image enhancement can be an impact or explicit goal of an event. Consider how the community's image has been altered as the result of an event, including how the media played a role in achieving this goal.

This illustrative list of factors to consider in gathering information on the impact of an event can potentially bring out positive and negative outcomes of the event, altering community dynamics, and these outcomes must be captured in any evaluation. Each area requires an evaluation method that seeks to do justice to the information it generates. The question must dictate the method.

Challenges in Conducting an Event Evaluation

Evaluation without corresponding methodological and practical challenges is similar to a fish living without water. It is impossible to separate the two and still be viable. Having challenges does not mean that the process is potentially flawed. Evaluation without challenges means that the process is simple and superfluous. The following are five factors that touch on the salient perspectives that must be taken into account when addressing the impact of celebratory events.

Motivation for Participation and Engagement

The subject of why individuals attend, and how the characteristics of attendees influence attendance and participation in events, has historically been the crux of many research and evaluation efforts. Motivation for attendance is complex and goes beyond simply seeking a disruption in daily life routines (Yoo, Lee, & Lee, 2015). Motivation for attending and expectations of an event are not uniform across all of those involved in a community celebration and this variety must be taken into account in evaluating an event.

A number of factors such as the presence of food, music, and availability of information on events are independent determinants of event satisfaction and loyalty (Novello & Fernandez, 2014). Attendees' sexual identity, age, level of acculturation, documented status, and gender all exert considerable influence on the nature and extent of event participation. Evaluation efforts must discern different ranges of expectations according to social and cultural factors. These factors, in turn, are further contextualized when grounded within an urban environment.

Why discern expectations to a high level of specificity? The answer is not complex but getting at the reasons certainly is (Benckendorff & Pearce, 2012). Understanding motivation provides important insights into the usefulness of specific community events and activities and how to improve community practice (Kaufman, Ozawa, & Shmueli, 2014). Targeting celebratory events is a central goal for communities, and having clarity regarding the intended audience translates into a focus on resources (marketing) to maximize their attendance.

Social Capital Outcomes over an Extended Period of Time

A variety of types of capital/assets have been associated with community celebratory events. Economic capital is one of those types, and some scholars would argue that there has been an overemphasis on this type. There is good reason for this attention (Carter & Zieren, 2012, 6): "Since the economic downturn, local governments have been faced with difficult budget decisions. Many local communities rely heavily on festivals and special events to generate spending and increase the influx of new money into the local economy." Carter and Zieren (2012) highlight the economic ripple of festivals and other events, increasing their significance, particularly for political and economic stakeholders.

Social capital also stands out in prominence by the coverage it has received (Onyx & Leonard, 2010). Capturing social capital necessitates the use of a protocol that allows measurement over an extended period of time and establishes the strengths of the relationships over time. A major challenge in measuring social capital, but particularly in relationships developed by participation in community events, is the longevity and significance of this capital (J.W. Smith, 2012). This requires a longitudinal evaluation to measure its lastingness.

Seven different types of community capital are outlined in this book, and they vary in strength and presence across urban communities and celebratory events. However, community capital takes on even greater significance in the case of marginalized urban communities that face daily social, economic, and political struggles.

User-Friendly

Evaluation is an activity that rarely is embraced with enthusiasm by staff and community, and faces challenges in finding wide acceptance; as result, evaluation is often relegated to professionals and academics. However, there is no disputing that if the evaluation of a celebratory event is to integrate community residents in prominent roles, evaluators must seek ways that result in the development of new approaches and methods that are user-friendly to individuals with minimal or no experience in conducting evaluations. User-friendly evaluation of community events will increase the chances that residents can effectively capture important nuances that increase the odds of the events being successful, and increase the likelihood of their participation and ownership of the outcomes of community events. This introduces a further challenge to an evaluation process that is often fraught with obstacles and tensions under the best of circumstances.

The transitory nature of event crowds, particularly at parades, and the need for obtaining views from different demographic groups, requires an evaluation that can capture this dynamic movement and unique groups' perspectives, and in a manner that provides sufficient depth, without seriously slowing the movement of participants in activities. This does not preclude follow-up over an extended period of time, which is important in order to understand the lasting power of participation. Evaluators must capture how participation influences attendees and the experience of volunteers, altering their perceptions and behaviors.

Undue Influence of Corporate Sponsorship

Fear of money undermining an event is well understood. The participation and influence of corporate sponsors (money) is a theme that will increase in importance as urban celebratory events take hold in the immediate future, and the staging and support of these events become a part of community practice (Gudelunas, 2011). In the United States, "corporate money" is associated with politics and elections. Its reach and influence cannot be restricted to this arena and must be a part of any discussion of community events, particularly those that have evolved from small events to large ones commanding media attention, expertise, and the support of key political stakeholders.

A key challenge for communities is how to tap corporate sponsorships without losing control over decision-making powers in staging their celebrations. Funding can come with strings attached as sponsors wield influence over decision making in planning the event, and this is particularly the case when sponsors are largely financially responsible for an event, and they seek to influence community perceptions of corporations with unsavory reputations. Community social work practitioners can help communities embrace diversification of funding sources as a goal to ensure that they are not compromised in staging their events and the political messages they convey. Thus, event sponsorship is a topic worthy of discussion, having practitioners help communities navigate these treacherous waters.

Segment Exclusion (Intentional or Unintentional Discriminatory Practices)

Segment exclusion refers to the systematic, intentional or unintentional, lack of engagement of select community groups because of their beliefs, characteristics, or abilities. In essence, it is the practice of discrimination. When applied to celebratory events, it means that there are groups that are undervalued and considered "unattractive" and are thereby judged as unworthy of participating as attendees or volunteers in these events.

The exclusion of groups is the ultimate irony because the excluded are by no means in a celebratory mood. Having a deep understanding of the segments that comprise a community, and how they can be affirmed, enhances community practice. A. Smith (2014, 260) comments on inclusion: "Nevertheless, staging events causes a series of intended and unintended consequences for public space provision. When events are ticketed, and where they require large installations, they exclude people symbolically and physically. Some legitimate arguments are used to justify event projects, but these are no consolation to those who feel their everyday spaces have been appropriated and/or violated." Sitting on the sidelines while there is a celebration is never a pleasant experience, representing the extreme opposite reaction of someone having fun and being a part of a positive lived experience.

The use of public space for celebratory events must be examined to maximize participation from all community segments. These events are interconnected with spatial configurations and everyday life (Lavrinec, 2013), with the potential to enhance a wide variety of community assets. Evaluators must confront the sensitive topic of who was not welcomed and how this message was conveyed to ensure that it was heard. The politics of inclusion/exclusion are bound to cause controversy, and community celebratory events will be the window through which this tension can be captured by practitioners.

Innovative Strategies for Evaluating Community Celebratory Events

Evaluating community celebratory events in a culturally and contextually affirming way is exciting and lends itself to highly innovative approaches to evaluation. This final section addresses this challenge and the advances necessary to surmount potential obstacles that evaluation places on participatory processes the search for approaches sensitive to the unique nature of events when community-centered.

The following five strategies will help uncover the nuances inherent in community celebratory events. These strategies emphasize capturing images associated with celebrations, and creating an understanding of the importance of events being inclusive of perspectives and opinions from a variety of population sectors, particularly ones that have historically not been sought and heard.

Visual Approaches

Participation in a celebratory event is a total sensory experience that results in a very special ambiance, with visual stimuli being an important dimension. However, the visual aspects of events are difficult to capture through reliance on the written word and reports. The emergence of visual ethnography ushered in an era of excitement and opened up areas for study through the use of photography (Matthews, 2014; Mitchell, 2011; Pauwels, 2011), including participatory video (Coffman, 2009; Daniels, 2012; Mitchell & de Lange, 2012).

These decisions are highly politicized with implications beyond the event itself. Video has potential for capturing the impact of an event, and in a manner that can be shared across wide audiences (Sarkissian, 2010), maximizing the narrative being shared. This is applicable because videos, digital in particular, can be altered to emphasize a narrative. A well-produced video is worth more than a thousand words (Malek, 2011). Digital forms also increase distribution options, reaching wider arenas.

Photography has been advocated for use in developing a nuanced understanding of events, and introduced the field of visual imagery as an innovative research and evaluation method (Malek, 2011). Photovoice can be used to evaluate celebratory events, and can offer a user-friendly approach: "Photovoice is a participatory action research strategy by which people create and discuss photographs as a means of catalyzing personal and community change" (Wang et al., 1998, 75). It is an emerging ethnographic research method with worldwide appeal (Delgado, 2015).

Photovoice can capture perceptions and sentiments through the use of narratives related to attendance at celebratory events through its emphasis on imagery and principles of participatory democracy (Nykiforuk, Vallianatos, & Nieuwendy, 2011). Just as importantly, photovoice must have community residents and attendees play a central role in gathering these photographs and accompanying narratives.

Balomenou and Garrod (2014) describe another use for photography as an important source for data that can be used in the planning of community events. The use of photography is not restricted to evaluators but can also be a way of involving volunteers, residents, and/or attendees in recording celebratory events for analysis that can also introduce fun and novelty (Becker, 2013). Vieira and Antunes (2014) encourage the use of photo-surveys to inform social planning projects, with implications for how community events unfold, which can also be used for evaluation and teaching.

Leonard and McKnight (2013) use photography (photo-elicitation) as a method that has particular appeal for youth, with equally important implications for capturing attendees' experiences of celebratory events. Buggenhagen (2014) highlights how photo albums can be a vehicle for capturing historical moments that can be shared with those who were not present or for future generations. These photographs can be put into event collections that can accommodate local goals (Mattivi et al., 2012).

Capturing and maintaining narratives of community celebratory events becomes an important goal for current and future generations, and visual methods can facilitate this goal. Cultural archives maintained and stored within communities in places that are easily accessible provide a mechanism for retrieval of information (collective memories) that would otherwise be forgotten and lost, and can be preserved for future generations (Bastian, 2013).

Maintaining a history of a community's celebratory events is a way of capturing history written by, and for, the community (both current and future residents). These narrative archives can be maintained in local libraries, schools, and key community organizations. These archives can consist of digital film, photographs, audio tapes, videotapes, newspaper clippings, journals, and other forms that are culturally based and relevant to a community.

Case Studies

Evaluation methods that lend themselves to the involvement of individuals with various skill-sets are always welcomed in community initiatives (Carlsen, Getz, & Soutar, 2000). Case studies help community practitioners grasp narratives that evaluators and researchers wish to convey, particularly when they wish to reach lay audiences (Yin, 2014). Use of case studies should not be surprising as the field of celebratory events attempts to broaden understanding of how such events change the hosting communities (Merrilees & Marles, 2011; Okech, 2011; Wong, Wu, & Cheng, 2014).

As the field evolves, case studies will assume a smaller role in knowledge creation although they will always have important contributions to make because of how they capture the context of events (Mehmetoglu, 2001). In the initial stages, case studies provide academics and community practitioners with lessons learned, and identify research methods that are user-friendly and locally grounded. Case studies provide contextual details that increase our understanding of the nuances that shaped an event, including insights into why certain segments of the community did not participate. Scholarly articles rarely have space to help readers understand all of the key social forces that may have shaped an event. Case studies facilitate conversations and exchanges.

A mixed-method approach introduces innovation and the use of unconventional approaches. For example, it is recommended that evaluators walk the parameters of the space where a fair or festival was held and the length of a parade route, in order to develop an appreciation and understanding of how geographical place plays a role in how a celebration unfolds and is received. Walking is one way of understanding surroundings that cannot possibly be understood by examining photographs or even looking at a map (Bairner, 2011).

This insight can easily fall under the realm of participant observation, or ethnography. However, insights garnered through this approach are not expensive or methodologically complex, with the potential for evaluators and researchers to team with residents in generating knowledge that helps contextualize and explain the community's celebratory events. The selection of the site for an event is one of the more important decisions a community can make, and evaluators must understand the community's reasoning in order to help other communities in their decision-making processes (McGurgan, Robson, & Samenfink, 2012).

Participatory Methods

Participatory approaches ensure that celebratory events meet the community's goals and expectations. Empowerment and principles of participatory democracy represent a central core of capacity enhancement and can guide the evaluation of an event, having community residents and event participants as members of an evaluation advisory committee to develop appropriate measures.

Madyaningrum and Sonn's (2011, 363) discussion of qualitative methods and the community art movement in Australia draws implications for how urban context and community participation has applicability to the role and function of how events are discussed in this book:

> We started the study with an assumption that the power of this kind of community art project is in the story performed in the play. We saw the story as an invitation for people to re-look at the history of the region and to reflect on the different ways of looking at the town's history – silenced and invisible versions of history. The findings, however, show that it is not the story that is central for the participants in the study. In the participants' accounts, the value and meaning of participating in the project is mostly associated with the process of actually being together with other people in their community, especially in the context where interactions across social groups are very limited. For the participants the value of this project lay in its ability to bring together diverse people in the community. For them, this togetherness is in itself sending a strong message, which perhaps is stronger than the messages embodied by the story performed in the play.

Exploring community residents' conceptions of what it means to "participate" in celebratory events also captures how local cultural factors shape their expectations and world views of what these events mean to them and their community. The act of participation takes on prominence and represents an essential element in how events have been conceptualized in this book.

Emerging Technologies

Emerging technological advances introduce innovative approaches to evaluating celebratory events and in a manner that encourages the involvement of many different groups, answering questions in a noninvasive and noncumbersome manner. The emergence of mobile positioning devices has opened up a new source of data to determine, for example, how far attendees might travel to an event and the potential of an event to attract out-of-town visitors, or its gravitational pull (Nilbe, Ahas, & Silm, 2014).

Social media increases in significance for marketing and evaluating celebrations, particularly events targeting younger segments of the communities and those most comfortable with these forms of communication. These data capture nuanced information in real time, adding depth to our understanding of attendees' experiences. Having attendees take photographs brings a visual dimension that can be combined with a corresponding narrative, as in photovoice (Delgado, 2015). This technology can, for example, facilitate the engagement of attendees with disabilities by pairing them with individuals without disabilities, bringing previously untapped perspectives.

Delphi Technique

The Delphi technique has been used to assess the impact of festivals (Kostopoulou, Vagionis, & Kourkouridis, 2013). This qualitative method has been in existence for approximately 40 years (Molnar & Kammerud, 1977), being in and out of favor over this period. Recently it has enjoyed a resurgence, providing event organizers with an opportunity to use findings throughout all facets of event planning, implementation, and evaluation. Communication technology has made the Delphi technique less expensive and cumbersome than when it originally was introduced using paper, pencil, and the US mail system.

This method has several distinct phases. The initial phase involves setting up a committee of members representing various important segments of a community. Key stakeholders can be enlisted as central figures in this method, with the size of the committee being flexible (ideally around 8 to 10 members so that the various segments are represented but the group is not so large that it becomes unmanageable). The Delphi technique brings sufficient flexibility to allow many different key interest groups to assess the impact of an event and at relatively low cost, over a relatively short period of time depending on the research goals.

Fifty to 100 community members are recruited to participate in the second phase, with each committee member recommending 10 participants, for example. The process is explained to these participants, and they are told that their responses will remain anonymous. They are asked to write a statement to respond to a question such as

"Why is a particular festival important for a community?" Their answers are sent back to the committee, which categorizes them into 20 groupings of corresponding statements.

These statements are then sent back to participants, and they are asked to rank order them as to their importance, and then they are sent back for tabulation. The committee devises a ranking from 1 to 20. This list is then sent out again, and participants are asked to rank order the items according to likelihood of a festival achieving the goals. These are then rank ordered by the committee, which will have two lists. The first list has the top 20 goals for a festival; the second will be a list of the easiest-to-harder goals to achieve. Festival organizers can then assess what are the most important goals and what are the most significant challenges in staging the festival. These results can be followed up with a series of key informant individuals to help further refine goals and strategies.

Evaluation Reports and Their Dissemination

It is appropriate to complete this evaluation chapter with a discussion of the final report and how best to conceptualize and disseminate its content to reach multiple audiences within and outside of the community. The issuing of an evaluation report can be a source of community celebration and a way of ensuring transparency and of furthering community engagement (LaFrance, Nichols, & Kirkhart, 2012).

A successful community celebratory event becomes obvious to all who worked on and attended the event. There is something about the "ambiance" that exudes success, and this must be captured in an evaluation and distributed, particularly in urban communities without a track record of producing "positive news" to the outside world. However, the voices of those residents of the community who did not work on or attend the event, for a wide variety of reasons, must not be overlooked.

Finding creative ways of disseminating evaluation reports in these circumstances takes on importance because a record of this achievement must be shared and saved for future generations, and it will be a valuable motivator for enlisting even greater civic engagement with future events. As an evaluation report's audience broadens, so does the impact of the social and potential results of the event itself (Kenny, 2011).

Evaluation reports are a vital part of an intervention, allowing for active community participation and enhancement of the participants' competencies (human capital). Their impact can go beyond issuing a report and actively engaging a community, creating memories and a basis for further actions. Creative ways of reporting and disseminating celebratory event reports also help communities across wide geographical boundaries learn from each other's experiences.

Conclusion

The importance of evaluation will increase in the future, and must do so if community celebratory events are to withstand the increased scrutiny associated with funding from institutional sources. As community events increase in frequency and significance, there will be a call for methods and techniques that are sensitive to these types of events, including modifying existing ones as well as creating new ones.

Evaluation is a potential minefield. Nevertheless, the rewards of an evaluation undertaken well can yield tremendous information and insights. Active and meaningful community involvement only heightens the rewards and challenges. Evaluators have a responsibility to evaluate an event and enhance resident competences in the process. The golden years of celebratory event evaluation are still ahead and the next decade will bring exciting new methodological advances and generate a better understanding of how to transform marginalized urban communities through the staging of community celebratory events.

4 Capacity Enhancement of Community Assets

The use of community celebratory events as prominent mechanisms for achieving social, cultural, political, and economic goals requires embracing a practice paradigm and framework that can guide social work practitioners in navigating the stages associated with the complexities of social interventions and integrate the community in the process. Scholars, too, benefit from embracing a paradigm and framework that identify and integrate the factors at work in the staging of celebratory events.

As addressed in chapter 1, a community capacity enhancement paradigm and its corresponding framework offers the greatest potential for use with celebratory events because of the emphasis on assets and resident participation in the staging of a community celebratory event. This chapter elaborates on the material covered in the introductory chapter, and also introduces a definition, set of values and principles, and the role of a framework in helping to shape celebratory events.

Theoretical grounding of this paradigm is essential to appreciate the factors that influence how it can unfold in "real life" practice with celebratory events. However, prior to describing the various stages and theoretical and political considerations in a community capacity enhancement paradigm, the reader must first be made aware of the social values that guide how community interventions based on this paradigm get manifested.

It is rare that an event's publication specifically identifies the values guiding the celebration, with the possible exception of economic values, although there may be many other implicit and important values, which are left to the reader to guess and identify (Mahadi, Hadi, & Sino, 2011). Yet values influence our world view, and that is the case with capacity enhancement and community celebratory events. We often take these values for granted, and rarely are they articulated, discussed, or even debated.

Evolution of Community Capacity Enhancement

All definitions related to emerging paradigms undergo changes as part of an evolutionary process. There are several different key elements in the following definition of *community capacity enhancement* (Kretzmann & McKnight, 1996, 1): "the range of approaches that work from the principle that a community can be built only by focusing on the strengths and capacities of the citizens and associations that call that community 'home.'" The emphasis on the centrality of the community's residents in crafting solutions and the importance of focusing on assets rather than problems stands out.

There are distinct values, as well as physical and sociopolitical elements inherent in a capacity-enhancement paradigm. Such a paradigm can help increase the likelihood that helping professionals can envision an active and meaningful role in using celebratory events as part of their practice (Traverso-Yepez et al., 2012). Wu and Pearce (2013b) applied an asset-based community-development paradigm to the development of tourism in Tibet, and illustrate how community practice and tourism can incorporate capacity enhancement.

The synergistic outcomes resulting from the embracing of a capacity-enhancement paradigm for use in guiding events can be significant in scope, with far-reaching social, economic, political, and cultural implications for marginalized urban communities (Getz, 2015; Ziakas, 2010). The emergence of place identity literature has introduced a perspective on public place and the role of social, cultural, and physical elements for aiding newcomer national identity formation (Main & Sandoval, 2015), which is often a theme of celebratory events.

Increasing attachment to urban place through the use of events adds an important dimension to community engagement or participation because of the importance of connectedness resulting from staging a community event (Raymond, Brown, & Weber, 2010). To achieve this engagement and sense of connectedness requires that celebratory events be carefully planned and implemented, with a clear sense of the goals to be achieved, an explicit paradigm, locally grounded, and active outreach to all major segments that comprise the community. The concept of "our" celebration is inclusive and gives true meaning to that word.

Community Capacity Enhancement Paradigm

A community capacity enhancement paradigm brings a distinct perspective on urban celebratory events. Altering a community's physical environment is one of the cornerstone goals of capacity-enhancement initiatives (Delgado, 1999b). When community events create shared welcoming public spaces for all members of the community, capacity enhancement can create positive social change and generate projects with profound meaning. Unfortunately, public spaces can also be spaces where exclusion and exploitation can transpire under the guise of affirming activities (Davilla, 2012). Events can exclude and exploit.

The spatial perspectives of urban events have become an important analytical dimension, as in the case of community organizations, in helping to understand who is motivated to attend and support celebratory events (Munro & Jordan, 2013). These considerations have an interactional (political) dimension, too. For example, who was pressured to attend, sponsor, or work on the celebrations? In essence, it is forced labor and counterproductive to democratic values.

Hayden (1995, 37), 20 years ago, addressed the role and influence of spatial context for appreciating festivals and parades, and his work fits well within an analytical and interactional perspective integral to capacity enhancement: "Festivals and parades also have to define cultural identity in spatial terms by staking out routes in the urban cultural landscape. Although their prominence is temporary they can be highly effective in claiming the symbolic importance of places. They intermix vernacular arts traditions (in their costumes, floats, music, dance, and performances) with spatial history (sites where they begin, march, and end)."

The artistic talent needed to design and create floats and other celebratory objects are a form of art reflecting cultural themes and local history (Hoefferle, 2012; Vanderwaeren, 2014). Hoefferle (2012) describes the role of community artists and floats in a US Labor Day Parade in Michigan's Upper Peninsula and how artists maintained local traditions, facilitating play and the development of social capital. The exposure of local artists also enhances economic capital by providing opportunities to be hired by local establishments (Correll, 2014).

Hayden's (1995) integration of community spatial context within celebrations shows the close interrelationship between space and events. Spatial politics, including the influence of urban street culture, bring an often-overlooked dimension to the study of community events (Pan, 2014). Lyck, Long, and Griege (2012) position events as vehicles for achieving community well-being in times of crisis by tapping (enhancing community capacity) culture as an asset.

A capacity-enhancement paradigm brings an added dimension to praxis and the need for social change to be a goal of celebratory events (Frostig, 2011, 50): "Arts activism is conceived in part as a fluid engagement with an idea and a process that joins personal motivations with social concerns while maintaining political currency." Social action against outside efforts to stage a community event is possible, bringing the community together to dictate what event can transpire within its borders (Hwang, Stewart, & Ko, 2012). It is not sufficient to stage an event; efforts must also be made to achieve positive social change or capacity enhancement that the residents of the community can embrace (Kohl-Arenas, Nateras, & Taylor, 2014). Events can be a manifestation of community activism (Collura & Christens, 2015; Lucas, 2014).

Local circumstances dictate how celebratory events unfold. A parade followed by a festival may be the most appropriate approach in one community but not recommended in communities that are highly fractured or underorganized. A fair or street party, however, provides for a smaller portion of the residents of a community to gather and celebrate, making it a more viable option, because of time, cost, and effort. In streets there is a high level of bonding social capital; however, a street party may not be conducive for bridging social capital because that is not its primary goal.

The scale of an event does not minimize these efforts and the goals they seek to accomplish. In essence, bigger does not necessarily translate into better. An assessment that takes interactional goals and factors into account is needed. A well-orchestrated community celebratory event, with proper media coverage, can accomplish important goals, regardless of size. Further, Ziakas and Costa (2011) and Welty Peachey and colleagues (2014), although specifically addressing sports events but applicable to other celebrations, argue that events must be leveraged for other community benefits.

The creation of urban green spaces, for example, creates opportunities for growing food and for engaging in physical exercise in a safe environment, as well as creating collective social memories, identities, and other forms of capital (Barthel, Parker, & Ernstson, 2015). These green spaces lend themselves to integrating newcomers from very different backgrounds, which often do not share similar languages, but might share an agricultural history and possibly celebratory customs surrounding harvests, for example. They also lend themselves to intergenerational projects (Delgado, 1999b). It is important to emphasize that they can result in spaces that are conducive for fairs and festivals, extending their functions beyond environmental beautification and food production.

Green space can be hard to find in cities and can be viewed from a variety of community assets perspectives. Human capital can be enhanced when new generations are introduced to the science and process of growing healthy food. Increased health derived from exercise and better nutrition, too, enhances human capital. Economic capital is generated through savings achieved in growing food and the funds obtained through the selling of food at farmers' markets. Cultural capital is enhanced when newcomer groups grow food that cannot be otherwise obtained locally. This food has symbolic meaning and plays a ritual role in connecting current situations to past experiences in their countries of origin and shared in celebratory events.

Physical capital is created when unusable land is converted to useful purposes and beautifies a community. Political capital is generated when garden plot shareholders come together and make decisions on the future of the green space. Finally, intangible capital is produced when residents are provided with a physical alteration to their environment that symbolizes hope and achievement and that can be a staging place for future celebratory events. In the case of green space, its benefits result in all forms of capital being enhanced.

The relationship between capacity enhancement and event staging is natural since this paradigm emphasizes cultural values and heritage as cornerstones of community-based interventions. Festivals can use heritage as an instrument of statecraft and as a tool for the assertion of grassroots political and economic agency (Henry & Foana'ota, 2015). Asset assessments guide how capacity enhancement unfolds. In this instance, community celebratory events are an asset to be assessed and a vehicle for further enhancing assets. This duality can be viewed as a challenge (Delgado & Humm-Delgado, 2013).

Values Underpinning the Capacity-Enhancement Paradigm

Values are the foundation that holds together social interventions. Values regulate social structures and human behavior, and values can be conceptualized as the criteria by which judgments are made about what can be considered legitimate actions or, in this case, worthy of a community celebration (Moeran & Pedersen, 2011). It is imperative that values be specifically addressed to ascertain their contributions to shaping a vision for practice, in this case, involving fairs, festivals, and parades.

Values share similarities with a cloud metaphor. It is possible to see clouds, describe them, track them across the sky, and even admire them. However, try putting your hands through a cloud and see how difficult it is to hold one. There is no denying that they exist. It is impossible to fully understand the role and function of community celebratory events without a corresponding understanding of how values shape the ways that attendees, community practitioners, and academics see these events. Context goes beyond space and place and the actors involved.

Much has been written about values in the helping professions, and one can easily be overwhelmed by this literature. Unfortunately, the event planning literature has not benefited from this type of attention. Goldblatt (1997) addresses the importance of core values and understanding how they shape events. Having value clarity provides event organizers with a broad "blueprint" of why to stage an event, and how it should benefit a community.

Core Values of Community Capacity Enhancement

There are a core set of values that are influential in guiding community capacity enhancement practice, although the reader is warned that this set is certainly subject to debate and represents one individual's understanding. The following seven values stand out in importance when applied to community practice and celebratory events. These values are presented in isolation from each other, yet this is a very artificial distinction. These values are not presented in any order of importance.

Social Justice

Social justice takes on great significance in a community capacity enhancement paradigm and particular relevance when discussing urban marginalized communities (Anguelovski, 2014). Inequality is associated with social justice and manifested in public places and spaces by restricting who can occupy these spaces (Ferreira, 2012). Social justice and cultural diversity can be considered "two sides of the same coin" (Ratts, 2011).

Harvey's (2009) book *Social Justice and the City* grounds this concept from a geographical perspective, illustrating how "a just distribution" stance helps academics and community practitioners further social justice. The quest for the "just city" provides a way of integrating social justice into transformative planning, in an urban context, including the role of celebratory events in transforming the community (Song, 2015).

Prilleltensky (2012) addresses social justice through a wellness lens, a concept that is finding saliency within a broader conceptualization because the various manifestations of justice (optimal, suboptimal, vulnerable, persisting injustice) translate into conditions that affect well-being. A state of wellness cannot be achieved if odds are stacked against meeting basic human needs and, to add insult to injury, being blamed for one's circumstances.

Sustainability and social justice, too, can be lodged within a community capacity enhancement paradigm, thus shaping how community practice gets operationalized (Larsen et al., 2014), and this can play a critical role in helping celebratory events to enjoy community-wide appeal. It can also be lodged within an urban grounding since context shapes the different dimensions that determine how an event can be viewed (Dempsey et al., 2011; Gilbert, 2014).

Hart (2014) focuses on a topic discussed earlier, namely, play, and how play is a basic human right for children and a fundamental aspect of their lives, drawing on the UN Convention on the Rights of the Child. Children's play is often trivialized by adults and we rarely seek children's views or involve children in decision making, violating their rights and making the subject of play a social justice issue. Youth views and their participation are rarely sought by adults, and this rights violation can lead to social action (Delgado, 2016; Delgado & Staples, 2008). A children's parade or festival, without involving children in the planning, is a social justice issue.

Balaceanu, Apostol, and Penu (2012) address the relationship between sustainability, celebratory events, and social justice, adding an important perspective on this association. Sustainability is not possible without social justice being central to how it is achieved in community practice, including celebratory events.

Celebratory events that can effectively tap community visions and support will achieve sustainability and thrive, while those that do not will simply vanish and not be missed. Other values are associated with sustainability, but in marginalized communities, social justice plays a central guiding role (Campbell, 2013), and nowhere more so than in situations where events redress a negative public attitude about an ethnic or racial group's history.

Community Participation

Community participation, or what is referred to as community engagement, has a long history and can be found in virtually all academic disciplines, and certainly in the helping professions, although it remains an elusive goal and much debated (Eversole, 2012). Community participation in human services in the United States saw its introduction during the Great Society Programs in the 1960s, and it has waxed and waned in significance ever since.

The concept has found its way into social research. Community-based research often has some degree of community participation, and nowhere is this more prominent than in community-based participatory research (CBPR) (Israel et al., 2008; Muhammad et al., 2014). Not surprisingly, CBPR is central to a capacity-enhancement paradigm and practice (Delgado & Humm-Delgado, 2013). Scerri and James (2010) advocate for community participation in the creation of social indicators of sustainability, an area that has historically been dominated by quantitative measures, but has implications for the evaluation of celebratory events. Qualitative indicators of sustainability introduce a missing dimension to the measurement and conventional indices of sustainability, and can be combined with a quantitative measure, fostering a comprehensive and nuanced perspective.

Quick and Feldman (2011) distinguish between participation and inclusion by emphasizing the degree to which decision making is shared. Parker and Doak (2012, 177) address the illusion of community participation, which has implications for the staging of celebratory events, and must be eschewed if this concept and value is to have meaning: "The concern being that efforts and examples where community is invoked or claimed to be engaged can often be facadist or aim at incorporation of sections of the community rather than genuine participation. This laden term invokes an imagined cohesiveness and mutuality on behalf of members or populations within set bounded areas such as streets, villages, towns, or for particular groups within localities."

It is easy to overlook the level and significance of community participation in the moment of attending an event. Le, Polonsky, and Arambewela (2015) address social inclusion and identify the following four key themes, with each tapping a particular manifestation of how to operationalize it: social connectedness, the link with home (original) culture, the link with "host" and other cultures, and inclusive initiatives.

Cunningham (2011) discusses *urban citizenship* and proposes it as a concept encompassing more than a right to vote, but also a right to seek empowerment and create empowering circumstances for residents who are suffering from social injustices. This concept brings together social justice, participation (activism), and empowerment, which follows.

Arts activism, too, can be an effective vehicle for addressing social justice (Newton, 2011), and events can certainly be an ideal backdrop for this activism to find roots (Bell, 2014).

Tisdall (2013) broaches this topic from a transformative or capacity-enhancement perspective emphasizing how participants, in this case children and youth, increase their competencies, confidence, and creativity through meaningful engagement. Enhancing human capital can become an integral aspect of engaging youth. Achieving "genuine" youth decision making requires that significant structural and belief barriers be overcome (O'Connor, 2013). Human capital is rarely applied to children and youth, yet this form of capital is quite appropriate for this age group and how their competencies are increased through engagement in celebratory events. Further, children and youth often form a significant portion of newcomer communities, making them a particularly attractive current and future asset.

Community participation and events are inseparable when embracing a capacity-enhancement paradigm. The level and type of participation influences all facets of the undertaking (Hulbin & Marzuki, 2012), and events cannot be community-centered if they do not reach out to all significant segments of the community. In essence, the question of when do we involve communities is inappropriate. A more appropriate question is, how will the community participate in the celebration and engage in decision making?

Participation must involve all segments of a community's population. Children and youth must not be passive in community festivals, and they can assume active and meaningful roles (Burch, 2014; Solís, Fernández & Alcalá, 2013). Giving voice and decision-making powers to those who are unaccustomed to being listened to, such as youth, is an important step in empowerment and the exercise of participatory democracy (Coser et al., 2014).

Empowerment

Any discussion of a social intervention that seeks to establish community ownership of the outcome will have empowerment as a central value. That is certainly the case when discussing this paradigm and community celebratory events. Empowerment must be contextualized to account for ecological nuances (Delgado, Jones, & Rohani, 2005; Lindeman, 2014). Although empowerment has universal appeal, it is dependent on local factors dictating how it can be maximized so that events have empowerment as a goal (Gibson & Dunbar-Hall, 2006).

Value-based intervention is a subject dependent on how it is defined and has evolved over time (Epstein, 2013, 1):

> Empowerment as a goal and a practice, that is, a method to achieve its goal, started out with actual political and social oppression in mind. Yet, over the decades since Freire and Fanon, who continued the radical political tradition, its target and meaning, has slowly been vulgarized to include just about any process that increases the capacity of the individual to manipulate his or her environment ... As the concept has become widened in contemporary empowerment practice to include the mundane and trivial, so too *has* it stretched the conditions of oppression to justify its attention to common situations of need and quietly excuse its near-uniform ineffectiveness.

Epstein is arguing that how this concept gets operationalized can result in its losing its central meaning and power.

When events value empowerment, it takes on various manifestations, including the creation of ways that maximize meaningful participation and enhance the competencies of participants. Empowerment and participation are not restricted to one particular group, and how they get conceptualized is influenced by who is doing the conceptualization.

Assets First

Language plays an instrumental role in shaping perceptions and thereby approaches to all forms of community practice, including involvement with celebratory events. How and why we label certain resources and activities requires serious thought to labeling and the concepts we use, and this is certainly the case with capacity enhancement and community celebratory events (Green & Haines, 2011). As practitioners and academics, we must think about the words we use in describing what we do.

Much can be learned from the disabilities rights movement and the emphasis on the importance of language in shaping opinions and practice. Unfortunately, for far too long we used terms such as "the blind," "the deaf," "the hard of hearing," and "the epileptic" to describe people with various forms of disabilities. Inserting the term "people" first, before listing the form of disability, places emphasis on the person rather than his or her physical or intellectual challenges. The same value must be incorporated into events, so careful attention must be paid to the use of language.

All communities possess assets and do not consist entirely of problems and needs, an important stance in identifying, enhancing, and mobilizing these assets in communities. It is much more than semantics and is a philosophical and practical way of reshifting how marginalized communities are viewed. In essence, a shift in paradigms brings a new vocabulary.

Asking the following question is a prevailing deficit way of thinking about urban marginalized communities: Were there any community assets tapped in the celebratory event? That is the wrong question. The correct question would be: What assets were tapped in the celebratory event and how were they enhanced or mobilized? The assumptions are fundamentally different. The former is not sure there are any assets, and the latter assumes that there are assets, the only questions being ascertaining what they are.

Cultural Competence/Humility

Cities are becoming more culturally diverse, and this demographic shift is reflected in cultural manifestations through art, music, and celebratory events. Cultural competence and humility is a value that permeates most helping and educational professions. Identifying and affirming cultural values whenever possible allows community practitioners to practice within a context that has authenticity for residents (Trinidad, 2012).

The transformation of urban spaces from underperforming or unsightly areas is due to the influence of their energy, drive, and creativity, which is often manifested in celebratory events (Foley, McGillivray, & McPherson, 2012). The naming of food, events, experiences (celebratory moments) is culturally based, providing a window into cultural influences on this process (Gabryś-Barker, 2014; Graham, 2013; M-Y Wu, 2014), explaining to outsiders the meaning of celebrations beyond having a "fun" time.

Ray (2004, 75) situates cities as places where diverse groups encounter each other and offer the potential to positively change society: "Cities have always been places where urbanities from many backgrounds rub shoulders and encounter 'the other.' But simply to applaud urban diversity or even to publicly recognize and celebrate a socially marginalized community does not in itself eliminate processes of exclusion. It also does not preclude, if unintentionally, fostering new ones. Landscapes of cultural pluralism and marginalization, as Centre-Sud [Montreal] robustly demonstrates, can co-exist in even the most open of urban public spaces." There are few urban places and spaces where various segments of society can come into contact with each other.

When celebratory events are community-centered, the likelihood that cultural competence and humility will be exercised is increased since context, of which culture is a prominent part, plays such an influential role with this value (Costanzo, 2012; Fowler et al., 2013). Cultures influence how celebratory events unfold as communities change and respond to social and demographic forces (Bowdin et al., 2012; English, 2011).

This value has inherent challenges in molding identify. How best to measure and operationalize cultural competence is integrally connected to the other values covered in this section. Nowhere is this more prevalent than in fairs, festivals, and parades through culturally informed themes, which can be thought of as an umbrella unifying the event image a community wishes to convey to a broader audience (Salem, Jones, & Morgan, 2004).

Reliance on Local and Self-Knowledge

The question of epistemology leads to answers that are deeply rooted in the values of the sponsoring groups. Thus, valuing local and self-knowledge translates into ways to tap this wisdom, and how it gets incorporated into all facets related to staging celebratory events. Local and self-knowledge are referred to by other terms, most notably *informal and experiential knowledge* (Sassen, 2013), shifting the power of knowledge from academics to the residents of a community. This value influences how community capacity enhancement unfolds (Delgado, Jones, & Rohani, 2005).

Contextual understanding of community celebratory events is not possible through an emphasis on educational-expertise. An academic's ability to bridge local knowledge and scholarly knowledge represents one of the foremost goals of community practice. Tapping local knowledge gives insight into the rationales, rituals, history, and why certain goals are emphasized over others. Reliance on local and self-knowledge is a value that permeates our understanding of how practitioners can assist a community in staging events and developing pathways to how this knowledge must be sought.

Community Investment

Celebratory practice entails investing time, energy, and financial resources to ensure a community benefits from staging an event. A value predicated on ensuring that external resources are brokered as an investment in a community's capacity, including support for and enhancement of indigenous leadership, is a value that, if actualized, has long-term implications.

Professional practitioners' expertise legitimacy allows the community to access resources that they normally would not have access to. Universities possess immense financial and nonfinancial resources that can be mobilized to assist communities in staging events. Their potential to contribute to celebrations must not be overlooked. Community residents with leadership potential can have access to technical assistance, workshops, and even scholarships to achieve university degrees with the hopes of having them return to their communities after graduation.

Principles

Why bother with principles if values are explicit? Principles are an extension of values and serve as a bridge between values and theory. They are much more specific than values, yet they are not that explicit that they are sufficient onto themselves. In essence, principles are far from being a cookbook recipe. They play functional roles guiding practitioners and scholars in all phases of event staging, but they require more extensive elaboration.

Brown and James (2004), in a rare publication on the role of principles shaping the planning of events, identify the following five principles that can guide celebratory events: scale, facilitating audience engagement; shape, clean and simple lines; focus, increasing the likelihood of the audience achieving event-related goals; timing, with activities unfolding according to schedule; and build, understanding the event curve. These principles address various elements (analytical and interactional) of staging an event, reflecting the complexity and multifaceted nature of events.

Delgado and Humm-Delgado's (2013) nine principles for community practitioners are modified here to illustrate the relationship between values and celebratory events, helping practitioners dictate how their roles can unfold:

1 A community has the will and ability to help plan and implement a celebratory event in a thoughtful and purposeful manner.
2 A community knows what type of celebratory event is best for itself, and this knowledge is manifested through self and informal sources.
3 Ownership of how best to assess internal assets for use in celebratory events rests within, rather than outside, the community.

4 Partnerships between the community's residents and community practitioners are the preferred route for any celebratory event, with the process of staging being planned with, rather than for the community.
5 Reliance and use of community assets in one area will translate into other assets and facets of the community, often creating a synergistic effect.
6 Celebratory events must not reinforce community biases and exclusionary relationships.
7 Celebratory event planning must occur according to predicted timetables reflective of the community's circumstances rather than a predetermined schedule based on what can typically be expected.
8 Evaluation results must be disseminated in a manner that reflects the culture and community-specific preferences for communication.
9 Celebratory events must maximize external investments and the leveraging of these resources for the broader community rather than for some subgroups.

These principles bridge values and practice. Practice principles are more specific than values, guiding practitioners through murky waters. The principles are not exhaustive and some, just like values, may take on greater prominence than others. Yet event practice without a clear understanding of the values and principles guiding it is neither purposeful nor strategic, limiting impact. When values and principles coincide between practitioners, organizations, and community, practice is easier. Significant differences can prove arduous and contentious, with celebrations suffering.

Framework of the Community-Enhancement Paradigm

Delgado (1999b) developed the following five-stage framework, with each stage having distinct analytical (theoretical) and interactional (political) dimensions and goals to guide how they unfold: assessment, planning, building support, implementation, and evaluation. This framework takes a developmental perspective by introducing stages in the carrying out of events. The analytical dimension of each stage allows practitioners to draw on an extensive array of theories and concepts to conceptualize tasks. Interactional aspects tap political considerations and actions necessary to facilitate the development of an event in a way that minimizes political obstacles and maximizes the achievement of goals.

It is best to conceptualize framework stages or phases as dynamic, with each stage building on the previous stage, allowing practitioners the option of refining their activities. Thus, although the initial stage is assessment, it does not mean that discovery occurs only during that stage. Similarly, it is important to emphasize that building relationships (cooperative agreements, collaboration) has an entire phase devoted to it because of its significance; this does not mean, however, that relationship development is restricted to this one stage since it occurs throughout a social intervention.

Conclusion

A community capacity enhancement paradigm is based on an explicit set of values and principles that can resonate across academic disciplines and professions. This paradigm consists of multiple stages, with integration of analytical-interactional factors and considerations, which, in turn, facilitate the use of this paradigm in staging celebratory events with a high level of community participation and facilitating integration of theory and local politics.

The promise of the community capacity enhancement paradigm is matched by challenges in identifying, mobilizing, and enhancing various types of capital (Raeburn et al., 2006). The relatively short history of using this paradigm does not diminish its potential for significantly altering a community's well-being. Further, this paradigm's emphasis on participatory democracy, local knowledge, social justice, assets first, and empowerment facilitates celebratory events being considered by helping professionals who would normally not think of these celebrations as viable options for community interventions.

SECTION 2

Community Celebratory Events

5 Fairs

It is fitting to start this section with a focus on fairs, which have not received extensive press coverage or sustained scholarly attention, even though they are ubiquitous in urban and rural areas and have long traditions throughout the world and can assume various forms, taking into account local social-cultural context and political factors (Calestani, 2013; Khaire, 2012; MacLeod & Johnstone, 2012; Young, 2012).

The lack of media and scholarly attention does not take away from the importance of fairs and only reinforces the need to shed light on our understanding of this form of community celebration. That includes broadening our understanding of the impact of fairs across multiple forms of community capital, and how intra- and inter-organizational supports are generated in the process of staging these events.

As scholarly attention increasingly turns to fairs, particularly when focused on community practice involving previously uninvolved disciplines, new insights will emerge with long-term implications for using fairs to achieve community-centered goals. The golden age of research on fairs is just around the corner, and it promises to bring knowledge and lessons that can be transferred to other forms of celebratory events.

Community fairs are associated with a wide range of special occasions that are often highly localized and facilitate the expression of a community's unique talents and history, with food (usually ethnic and cultural) seemingly always being an integral part of these events. Fairs can also have various rides and attractions, including individual and team competitions, which further increase participation and community ownership. Fairs are rarely focused on one goal and often can address multiple community goals and assets, including those that are related to human services. This can increase their attractiveness in community practice but also brings challenges associated with multiple agendas, including evaluating the outcomes.

Fairs can be staged in urban locales of various sizes, taking into account a variety of considerations, challenges, and meanings. They can be quite large and take place in fairgrounds covering several acres, or limited in size to unfold in narrow confines such as a closed-off street or in a building. Here fairs differ from festivals, which tend to be of considerable size, and parades, which require a large, although constrained, geographical area. This physical flexibility allows communities to construct fairs to meet highly localized needs, budgets, and even seasonal conditions.

Definition of Fair

The following detailed description of what a community fair is includes a definition and a range of goals for holding a fair that can have wide appeal (Better Evaluation, 2005, 1):

> A community fair is an event organized within the community with the aim of providing information about the project and raising the awareness of relevant issues. It can include a wide range of activities that cater to a variety of different people and include such things as sausage sizzles, rides and activities for children, young people's activities and events of interest to adults.
>
> Community fairs can be useful for stakeholder engagement in that they provide an enjoyable venue that can draw a crowd of all ages and backgrounds. By using a variety of activities and events of interest to inform and engage a broad range of people a community fair can be a good way to raise awareness about an issue or proposal. The events incorporated within community fairs, if focused on certain issues, will encourage public participation and raise awareness on this basis. They can also provide a venue for collecting contact details and getting signatories to any submissions or alternate proposals.

The types and scopes of fairs, according to the Better Evaluation definition, are only limited by our imagination and can incorporate multiple goals, bringing flexibility to allow local circumstances to dictate who an urban fair reaches and how it unfolds.

A definition of the county fair as envisioned in the United States provides similar and added dimensions to the above general definition. Marsden (2010) argues that the primary goal of a county fair is to showcase the strengths and assets of community youth by using education, exhibits, livestock contests, and other contests, projects, and activities which youth participate in. In addition, country fairs illustrate how agriculture, industry, and businesses represent essential foundations of a community. Rural fairs can reach across age groups with age-specific activities that are unifying and appealing.

There is flexibility in how communities can stage fairs. There is no one definition, time frame, model, or focus for community fairs, and in similar fashion to other events covered in this book, fairs bring flexibility in rewards and challenges. World fairs and expositions often come to mind when the word "fairs" is mentioned (Greenhaigh, 2011); however, because of their size and the required financial and political commitments, world fairs are far and few between.

At the opposite end of the continuum are fairs sponsored by religious institutions, schools, and human service organizations. These types of fairs reflect a heavy community presence, and to succeed, they must ensure that they reflect local priorities. Local organizations such as civic organizations and hometown clubs, too, can sponsor fairs. The focus of this chapter is on urban community fairs rather than their mega-fairs counterparts, since they necessitate a different level of funding, logistics, and enlistment of major political and business stakeholders.

Parameters and Range of Types of Fairs

One of the first tasks practitioners and academics face is determining the scope and parameters of a celebratory event and the types that can be classified as falling into a specific category of events. Categorization helps generate an understanding of the immensity of the tasks involved in staging a fair that can reach out to all segments of the population. Discussion of types of fairs is of interest to more than just academics, helping to guide communities on options for meeting their local goals.

Any discussion of the parameters or range of types of fairs uncovers numerous kinds, including fairs with a focus on celebrating talent (Lövheim, 2014); agriculture (Sumner, Mair, & Nelson, 2010); auction (C.W. Smith, 2011); art (Moore, 2011; Thompson, 2011); music (Klein, 2011); careers (Gordon, Adler, & Scott-Halsell, 2014); country (Lampel, 2011); education (Gębarowski, 2012); science (McComas, 2011); trade (Bathelt & Spigel, 2012; Entwistle & Rocamora, 2011; Maskell, 2014); health (Knight, 2014); renaissance (Diehl & Donnelly, 2011); science, religion, colleges, crafts, or books (Moeran, 2011a); and vendors (Bray, 2011). This illustrates the wide appeal of this form of community event.

It is reasonable to ask whether there is any subject that cannot be used to create a themed fair. If there is not, then the evolution of urban fairs is still in its infancy with potential for individual and collective transformation.

The evolution of fairs can be understood through a closer examination of the themes associated with them. Hampton and Licona (2013) examine elementary- and middle-school children's science fairs in a Texas Mexican-American community near the Mexican border, as an example of the range of fairs. Conklin-Ginop and colleagues (2011) report on fairs used to introduce Latino youth to carnival drum and dance traditions, rarely practiced by these urban youth, and how these fairs then provide a public forum for them to show their newly acquired talents to neighbors, friends, and other residents. Prisons, too, have been sites for fairs; Oswald (2005), for example, looks at fairs related to employment, demonstrating that job fairs can be put on within totally institutional communities.

In the United States, the image of the county fair is integral to the national identity, with practically every resident being able to conjure up an image of this form of fair; an estimated 2,500 county fairs are held annually in the United States (Marsden, 2010). Street fairs are associated quite closely with cities, and they provide spaces for cultural groups through the production of traditional festivities and the sale of ethnic street food (Barron, 2011; Bueno & Milanese, 2012). The street as both place and space is well recognized in the social sciences.

These types of fairs can occur within a confined and small geographical area of a community, and it is not unusual to have multiple fairs occurring simultaneously, with the potential to foster civic engagement and interactions. The frequency and scope of street fairs can be significant (M.L. Jones, 2014), but this topic has not received much attention in the professional literature, particularly when compared with the attention given to county fairs. Mason (1996, 301) addresses the role and importance of street fairs and how urban space and place lend themselves to these types of events by stressing how street fairs in the United States are a form of public celebration that combines celebration with business, and also a form of cultural exhibition.

Street fairs are found throughout the world (Bueno & Milanese, 2012; Zinkhan, Fontenelle, & Balazs, 1999), sharing many similarities with the other types of fairs (Ibrahim & Sidani, 2014). Human service organizations can conduct outreach and screenings, with food, music, and performing acts often being integral parts of these fairs.

O'Sullivan and Jackson (2002) classify fairs into the following three broad categories based on their size and goals: (1) home-grown fairs, which are primarily intended for locals, are small in size, and focus on cultural and entertainment goals; (2) tourist-tempter fairs, which are medium-sized and focus on local economic development, seeking to bring in visitors; and (3) big-bang fairs, which are large in size, with a focus on large-scale economic development and the maximization of tourism dollars and major publicity for the sponsoring organizations. This categorization fails to capture the nuances associated with fairs, but nevertheless can still help practitioners classify them based on crowd expectations and economic impact.

Fairs embrace parameters concerning goals, sponsorships, and attendance that are different from their often majestic festival and parade counterparts. Fairs, more so than festivals and parades, tend to be smaller in scale and attendance expectations, and they have a higher probability to be used by helping professionals to reach consumers, as in the case of health fairs (Burron & Chapman, 2011; Cupertino et al., 2011; Landy, Gorin, & O'Connell, 2011; Murray et al., 2014; Padilla et al., 2010).

Festivals and parades rarely have human service goals, although there are no inherent reasons for this. Particular community settings rely on fairs for achieving a variety of goals related to health, education, welfare, and the engagement of the people living in the community (Seo, 2011). Schools and health facilities, however, stand out because of their high prominence (Main & Velovis, 2010).

Health fairs are popular vehicles for reaching large segments of the population and attracting support for community organizations. This appeal takes on greater significance when addressing newcomer groups for whom English is not the primary language, for health is a vital segment of human capital (Delgado & Humm-Delgado, 2013). Fairs are a culturally specific extension of their primary organizational mission (Burron & Chapman, 2011; Lucky et al., 2011; Murray et al., 2014).

Fairs encompass and enhance the same types of capital/assets as found in festivals and parades, including social, cultural, and economic, the three most frequent forms of capital associated with celebratory events. Fairs have been referred to as themed community spaces with cultural and political meanings (Brannstrom & Brandao, 2012). Art fairs, for example, represent an important outlet for the selling of locally produced art, which often captures local themes (Yogev & Grund, 2012), and they are thus a source of economic and cultural capital in these communities. Further, art fairs can easily accommodate various forms of art and the art created by different age groups or demographic groups. Art by people with disabilities, for example, can be sold at fairs (Barnes, 2003; Maclagan, 2010; Swain et al., 2013).

The selling of crafts, food, and other merchandise such as heritage-related items at fairs takes on greater importance in situations where entrepreneurship is otherwise not possible outside of the immediate community (Caust & Glow, 2011; Maclean, 2010; Rong-Da Liang et al., 2013). Agricultural fairs can involve more than the selling of agricultural products and also encompass learning and outreach activities (Darian-Smith, 2011; Lillywhite, Simonsen, & Wilson, 2013). Fairs can provide an accessible marketplace for locally grown agriculture, and in doing so increase place attachment (Dunlap, Harmon, & Kyle, 2013). Their commercial potential, particularly when focused on enlisting local merchants, brings economic capital to fairs.

Fairs have a multifaceted impact on communities and the economic benefits cannot be ignored. Both urban and rural fairs provide marketplaces for the selling of food, local products, and produce (Aslimoski & Gerasimoski, 2012; Larsen, 2012; Moeran, 2011b). Any analysis of this capital takes into account the formal and informal economy since many local residents may be involved in the creation and marketing of these items. Taking the informal economy into account increases the challenges for evaluation since this sector may be very suspicious of outsiders. In these situations, it will be critical for evaluators to have sponsors (individuals well respected within the community) to help them gain access.

When art fairs involve the painting of murals, they take on physical and intangible forms of capital creation, too. The cultural, social, and political capital that they generate serves to affirm and connect community groups sponsoring these fairs. Fairs can be sites where investment in human capital occurs (Chapple & Jackson, 2010). Lauver (2011) reports on the use of a health fair as a forum for attracting young children's interest in pursuing a nursing education and career. Health fairs, too, can serve as service-learning civic engagement sites to enhance human capital (Kolomer, Quinn, & Steele, 2010).

Service-learning is an effective mechanism for community development, bringing potential for use in celebratory events (Geller, Zuckerman, & Seidel, 2014). Increasing participant competencies opens up opportunities for engaging community residents based on their interests and enhancing human capital in the process. Participants can also receive high school and college credit by formalizing a learning contract, further opening up possibilities for advancement and possible careers in event planning. It is important to bear in mind that human capital includes health as a key element (Delgado & Humm-Delgado, 2013).

Rewards and Benefits of Fairs

Fairs bring rewards and benefits for the communities staging them. Fairs can serve as an upbeat vehicle for solidifying in-groups and out-groups (Moeran & Pedersen, 2011). Fairs can bring flexibility for communities taking into account size and complexity, allowing local circumstances to dictate how they unfold and the extent to which they require external funding and corporate sponsorships (Gudelunas, 2011). Not having to rely on extensive external funding helps local communities to stage a fair without having their goals compromised by sponsors and being encumbered by outside sources.

Furthermore, fairs can take place indoors and outdoors, unlike parades and festivals. This allows for fairs to be held during the winter and not to be subject to the vicissitudes associated with inclement weather. Fairs can be venues for incorporating the latest concepts and trends, such as the green movement, sustainability, civic engagement, and interagency collaboration (Ball & Wanitshka, 2014; Ezeonwu & Berkowitz, 2014; Gough & Longhurst, 2014). Community events must not harm the environment and the green movement has become part of fairs and other celebratory events (Dickson & Arcodia, 2010).

Researchers and practitioners try to find creative ways of informing educators and human service providers of how culture gets manifested in urban communities, and ways that service-learning and civic engagement, too, can be used to provide a service and enhance the competencies of those providing the service (Amerson, 2010; Billig, 2012; Delgado, 2016; Stoecker, Tryon, & Hilgendorf, 2009). Here a word of caution is in order for using fairs for learning about diverse cultures: an examination of fairs can be a creative way of enhancing a knowledge base on communities, but it can also reinforce stereotypes if the research is improperly undertaken and not well grounded (Maguth & Yamaguchi, 2010).

Urban ambiance represents a benefit of hosting fairs, with fun and play associated with this event. Fairs also provide sufficient intimacy to host booths offering learning opportunities and outreach to attendees. Fairs lend themselves to involving various age groups, including children, with responsibilities being tailored to developmental competencies and needs (Hart, 2013). These enclosed areas facilitate supervision and support of very young children particularly when they are held in settings that are children and youth-centered.

Fairs have spurred innovative strategies for reaching mass audiences. Mobile technology has opened up new arenas for deployment at fairs and other crowd-intense events (Abebe et al., 2013). Surveillance and tracking of event crowds can be controversial when no effort is made to inform attendees that this is transpiring, as happened with Boston's 2014 Calling Music Festival (Ramos, 2014). In this case, "situation awareness" software developed by IBM was used. Crowd-sensing technology is relatively inexpensive but limited in answering quantitative questions on crowd size and movement. Nevertheless, this perspective is often missed, particularly in real time. Social media allows these moments to be shared, if warranted, with those who cannot attend in real time.

Questions pertaining to geographical sites are not restricted to professional geographers and should be of concern to both residents and community practitioners (Devine, Moss, & Walmsley, 2014). The geographically focused nature of fairs facilitates the gathering of information on how the site influences outcomes, with similar considerations to individual events and booths at the fair. Location is important when considering real estate, and this consideration also applies to the fair as a whole and within the fairgrounds themselves (Everts, Lahr-Kurten, & Watson, 2011; Saayman & Saayman, 2006). Urban fair locations must be accessible by public transportation or within walking distance and be considered "safe" psychologically, socially, and physically.

Fairs can be used to reach and influence participants to enter certain professions, for example, in the case of veterinary medicine and state fairs in the United States (Laflin & Anderson, 2010). Youth attendees have the opportunity to witness firsthand the potential of a career in this field. Other professions, too, use fairs for recruitment. Fairs, particularly when catering to newcomer populations, can be sites for new discoveries that can help attendees with various health conditions. Leitão and colleagues (2012) report on a unique study of open-air fairs in Rio de Janeiro, Brazil, for the discovery of medicinal plants (ethnobotany and ethnopharmacology) to treat tuberculosis, and addressing the role of cultural, economic, and human capital development. Other forms of discovery are possible.

Tensions and Challenges of Fairs

Fairs may appear to be stress free and without significant challenges, and this conclusion is faulty. For example, fairs are often associated with animals of various kinds, allowing participants, and particularly children, to interact with them and thereby adding a dimension not found in parades and most festivals. Access to various kinds of animals takes on prominence in the case of urban children and youth who usually have limited access to them. Disease outbreaks related to animal-human interactions are not uncommon (Dressler et al., 2012; Marler, 2011).

Fairs and food are closely tied (Hartel & Hartel, 2014), raising the potential of food-related illnesses (Lee, Almanza, & Nelson, 2010; Morgan, 2010). Addressing food-borne illnesses at fairs, festivals, and farmers' markets is complicated because inspection at these types of events is handled differently than at restaurants, calling for specialized training for food handlers and the issuance of inspection guidelines for temporary food services that should differ from inspection guidelines for permanent food service establishments (Choi & Almanza, 2012). Where these vendors are undocumented, they may be reluctant to undergo specialized training by government officials for fear of arrest and deportation.

Nevertheless, culturally, some ethnic and racial groups have a close association between celebratory events and animals, and animals are part of a long-standing narrative that connects these groups with cultural heritage and their countries of origin (Barua, 2014; Yeh, 2013). The presence of animals and other attractions can play a positive role in increasing attendance (Lillywhite, Simonsen, & Acharya, 2013), but also bring challenges for protecting participants and animals alike, particularly when there may be attendees who have allergies that they are not even aware of; this can raise serious medical issues. Fairs must contend with food safety (Boo & Chan, 2014), as well as other potential mishaps associated with crowds.

Cornwell and Warburton (2014) address the under-researched topic of how work hours impact civic engagement, with considerations for community fairs and other events. They draw conclusions for newcomer urban communities, where individuals may well be involved in working unusual shifts, thereby reducing the potential of involvement in activities structured around conventional work hours and days.

In communities with high numbers of newcomers with low levels of formal education, the possibility of working multiple jobs and having untypical work schedules has implications for when fairs can be held. Local circumstances dictate what is considered a conventional or unconventional fair schedule, and this may result in tensions between different segments of the population that may feel that an unconventional schedule will detract from the central goals of the celebrations.

Fairs must remain current to have viability, but this, too, brings challenges and tensions, as happened with Denver's county fair, which became the first fair in the United States to allow recreational marijuana use (Colorado made consumption of recreational marijuana legal in 2014). In 2014, Denver introduced a "Pot Pavilion," where "blue ribbon" contests were conducted involving photos of bongs, plants, edible items (such as brownies), hemp clothes, and even a joint (oregano) speed-rolling event (Wyatt, 2014, A6): "Organizers say that the marijuana categories this year ... add a fun twist on Denver's already quirky county fair, which includes a drag-queen pageant and a contest for dioramas made with Peeps candles." Incidentally, the edible treats were judged earlier in the month. Needless to say, this entry has resulted in differences of opinion.

Evaluation and Research Challenges of Fairs

Although fairs can be small with limited goals, evaluation is no less important with these events since "big" is not necessarily "better." The evaluation of community fairs brings important insights and provides opportunities for residents to play a meaningful role in planning and carrying out evaluation. Evaluating community fairs shares many of the same potential rewards and challenges associated with other community celebratory events, but also some unique ones, too.

The intimate nature of fairs, particularly when compared with other larger and more elaborate events, enables a greater understanding of social relationships (social capital) and how they are altered or enhanced through attendance at or work on celebratory community events. The nature, scope, and goals of fairs when introducing human services calls for the use of innovative methods to buttress those methods that are more conventional, such as surveys.

The development of social capital is a central feature of fairs and plays a central role in any evaluation. However, its popularity as a concept is only matched by the difficulty in capturing this measure in a comprehensive manner. Poortinga (2012) draws an important connection between social capital and community resilience, but also addresses the complexity of this capital because of the various ways it can be manifested.

Murray and colleagues (2014) specifically address health fairs and the importance of systematically tracking attendees to evaluate the effectiveness of fair-based interventions. The use of key informants taps qualitative data and is not expensive to carry out in evaluations (Escoffery et al., 2014). Further, these interviews can be conducted by community residents if they are properly trained and supported. It is important that criteria for selecting key informants or interview participants be made explicit since one person's expert may not be someone else's expert. Fortunately, the power and reach of social media provides evaluators with instruments to help capture real-time experiences.

Community Practice Implications and Fairs

It would be a rare curriculum where the staging of fairs is either a course or part of a course or lecture series. Celebratory events, at least in social work, may get mentioned in a curriculum but are not systematically addressed. Yet, in similar fashion to other community events, we learn about fairs as attendees, or when organizations enlist our help in staging them. It would be a rare community practitioner who was schooled on celebratory events before entering practice.

Community practice can be involved throughout all phases of a fair (Matheson & Tinsley, 2014), and there are plenty of opportunities for engaging our services. Community practitioners are not expected to have equal knowledge and competences (analytical and interactional) in all phases and facets of celebratory events. They can have areas of particular expertise and comfort level. The potential for relative intimacy at community fairs, both street fairs and the broader community type, fosters involving all age groups, and practitioners are in an excellent position to develop outreach efforts to involve children and youth (Dodd et al., 2006). This can be accomplished through partnerships with local day-care centers and schools.

Art projects can be accomplished during the time children and youth are in these establishments, as can service-learning projects. Class projects bring a collective experience that is powerful in producing collective learning and memories. Youth in colleges and universities, too, can benefit from engagement in community fairs. Service-learning and social justice can come together in service through fairs and other celebratory events (Amerson, 2010; Delgado, 2016; Donaldson & Daughtery, 2011).

Fairs, again due to their size, bring potential for close social interactions, and their focus on a prescribed urban geographical area such as a street, park, or playground facilitate the engagement of local human and educational organizations in service to the community, particularly where these institutions are also located within the community; this is also highly applicable for those organizations outside of the community that do not have existing reputations to draw on.

Community practitioners can develop collaborative agreements involving these organizations, helping to determine what services are considered a priority by the community's residents. Prioritizing which services should be allowed into the fair is important; residents can easily become overwhelmed by a plethora of organizations descending onto a fair because it is an "easy" or "cheap" way of reaching large underserved or high-risk groups.

Human service and educational organizations often see fairs as a venue (culturally, geographically, and psychologically) for engaging hard-to-reach segments of urban populations. Thus, there is a need for community practitioners to help design working agreements and even help the community staging a fair to determine the number of organizations and their locations within the fairgrounds. Site considerations are not restricted to where a fair will be held, but also apply to where events and booths are located within fairs, with some sites being considered prime because of their visibility and accessibility.

The role of social broker stands out for community practitioners and taps key analytical and interactional competencies. There are a lot of negotiations, permits, and paperwork required to stage a community fair, necessitating a knowledge of bureaucracies and time to obtain these permits. Finally, community practitioners can serve as media contacts or media consultants, although these competencies are rarely addressed in most graduate school educational programs – certainly not in social work (McKay & Williams, 2010).

Conclusion

The paucity of scholarly literature on fairs translates into a need for more work to increase the knowledge base to equal that for the two other more popular celebratory events, festivals and parades. This bodes well for scholars interested in this topic. Fairs are complex undertakings, bringing their own particular challenges because on the surface they appear simple in design and staging, and when everything goes well, no one really thinks about the challenges in pulling them off. The level of coordination, participation, and hard work associated with fairs necessitates the extensive use of local partnerships and cooperative agreements. Fairs are localized vehicles with their own goals and logistics.

This chapter identifies a variety of fairs that encompass a range of goals and local purposes. Some fairs focus on cultural heritage themes, while others emphasize educational and human service goals. Regardless of themes, all fairs are community-centered, necessitating the active participation of residents, local officials, and stakeholders in staging them.

Fairs have the potential to enhance all assets, in similar fashion to festivals and parades, although their focus on specific assets is not unusual, as in the case of health fairs and human capital. Human capital goes beyond education and competencies and can encompass health. Lastly, there are numerous rewards, benefits, and corresponding challenges facing community practitioners who hope to harness the power and influence of community fairs.

6 Festivals

Who does not like a festival? Most of us have positive childhood memories of fun activities, laughter, food, and companionship when discussing our experiences with attending festivals. Not surprisingly, festivals are considered one of the most important aspects of the tourism industry (Thomas & Kim, 2011). As a result, it is impossible to discuss festivals without discussing money and tourism.

Festivals, or fiestas in Spanish-speaking countries, in similar fashion to social protests, can be manifestations of communal values (Belghazi, 2006; Calestani, 2013). Festivals celebrate place and space (Peters, 2014) and are common but complex phenomena, integral to all geographical settings, and bound to elicit positive reactions from host communities and attendees alike (Cudny, 2014; Ferdinand & Williams, 2013; Gibson et al., 2011; Tull, 2012). These events provide opportunities to influence stakeholder and broader community perceptions (Buch, Milne, & Dickson, 2011).

Discussion of globalization has generally been restricted to business, communication, or technology, and these interrelated and interdependent fields across the world. Community celebratory events, too, must be discussed in this context because they create a bridge between old and new worlds for the transplanted. Festivals can unite communities in scope and prestige, as well as cross national and international boundaries (Elias-Varotsis, 2006; Magazine, 2011; Packman, 2012). Sporting events have been seen as achieving these goals, too (Schulenkorf & Edwards, 2012; Schulenkorf, Thomson, & Schlenker, 2011).

Getz's (2010) review of the literature on festivals sets a context for appreciating the history of scholarship on this subject, and why it is relatively short and traced back to the early 1970s, although that period witnessed but scant scholarship on festivals. The early 1990s are considered the beginning of serious scholarship on this topic, with the turn of the millennium reflecting the most significant research and scholarship on festivals.

The Mexican holiday El Cinco de Mayo, marking Mexico's victory over France on that day in 1867, is probably the most widely known Latino holiday in the United States. Oddly, the holiday does not engender the same fervor in Mexico. Hayes-Bautista (2012) attributes the popularity of El Cinco de Mayo to states with sizable Mexican populations, and traces the forces that support this holiday and the influence and evolution of various socioeconomic, sociopolitical, and sociocultural interests involved with it. Initially, this was an event that fed nostalgic feelings for the old country (cultural capital), political activism (political capital) during the 1960s and 1970s, and commercial interests (economic capital) during the 1980s and continuing into the present. Evolution is an integral part of community festivals (Isaac-Flavien, 2013).

Definition of Festival

Defining *festival* can prove challenging and may indicate a field that is still evolving, even though festivals can be traced back thousands of years. The following definition represents a particular themed perspective: a *festival* is a "themed public celebration that involves tourism, leisure and cultural opportunities such as shows, dance, film, music, food, visual arts, crafts, harvest celebrations, sporting events, rituals and agricultural products" (Lee & Hsu, 2013, 18). Festivals can be called by many different names – most notably, carnivals or fiestas.

Parameters and Range of Types of Festivals

Festivals are thematic and capture sentiments or historical events. Lee and Hsu's (2013) definition highlights how the special interests and shared values and experiences of attendees shape their reasons for attending, a complex undertaking, even under the best of circumstances. This section on parameters and types of festivals illustrates their multifaceted forms and why they are attractive for urban communities.

Festivals know no geographical boundaries and can consist of varied themes and sizes, and thus they can be sponsored by any organized group. Zhu (2012), for example, examines Canada's national holidays and the accompanying "minority" festivals from 1867 to the early 2000s and concludes that these events serve as indicators of nation building, highlighting the outcomes of decolonization. Event tensions and debates reflect national ideology, necessitating a sociopolitical and sociohistorical perspective to understand their present-day significance and evolution (Newman, 1997).

Three focal areas of festival literature can be found leading up to 2010: the roles, meanings, and impact of festivals in society and culture; festival tourism; and festival management. Sassatelli (2011) suggests that the scholarly literature has suffered through an overreliance on impact evaluation, management issues, and the relationships between growing professionalism, commercialization, and the popular success of these events – thereby missing important elements.

Each scholarly area emphasizes a unique festival perspective, but it is important not to lose sight of the overall contributions of festivals to urban life. Festivals can focus on a wide range of themes, including food (Kim, Duncan, & Jai, 2014) and destination branding to increase tourism traffic and revenues (Andersson, Getz, & Mykletun, 2013b; Lee & Arcodia, 2011; Ryan, 2006). Fairs, in contrast, are rarely driven with destination branding as a goal.

Giorgi, Sassatelli, and Delanty (2011) argue that the concept of *public sphere* has concentrated too much on studying strong institutions, such as religious and political, and on identity formation. They advocate for a greater use of "public culture" as a means of understanding cultural significance, including the study of festivals (Sassatelli, 2011).

Any examination of the parameters and types of festivals will uncover a rich tradition and landscape highlighting flexibility and purpose for staging these events. This range of possibilities will prove very attractive in community practice because it facilitates local goals and considerations dictating the type of festival that will be staged. Although this chapter focuses on community-centered festivals, there are festivals that travel from one location to another, bringing a dimension that is beyond the scope of this book (Bochenek, 2013).

Festivals are not often associated with social justice themes. Bogad (2010) focuses on carnivals as oppositional events, however, and shows how they can be. Lucas (2014) ties community organizing to capitalizing on festivals. Festivals can be used to address stigma, as when a national mental health arts festival was mounted targeting the general public (Quinn et al., 2011). There are film festivals devoted to social activism (Calley Jones & Mair, 2014; Iordanova & Torchin, 2012; Stover, 2013).

Janiskee (1995) undertook a novel research effort approximately 20 years ago and tallied the number of festivals in the United States over a one-year period. There were an estimated 20,000 festivals across the country, with the prime season for them being May through October, with a peak in July and a secondary peak in September. This number is dated, however, and a similar study will witness a dramatic explosion of festivals in this and other countries (Dinaburgskava & Ekner, 2010). In 2012, China had over 5,000 festivals, and that number, too, is becoming dated (Congeong, 2014).

In 2009 the US National Endowment for the Arts (2010, NEA) undertook a national study of outdoor arts festivals and found eight major self-reported types: arts and crafts, child and family, multidisciplinary, music, performing arts, racial and ethnic, theater, and visual arts festivals. Music, visual arts, and crafts are the most popular types of festivals. Categorized according to artistic discipline and popularity, the NEA findings show clear preferences for the types of festivals the communities organized (with many communities organizing multiple types of festivals): music festivals (in 81% of communities), visual arts and crafts festivals (67%), dance festivals (42%), folk and traditional arts festivals (41%), theater festivals (22%), literary festivals (19%), other festivals (12%), and film festivals (4%).

Festivals are ubiquitous and explore a wide variety of subjects such as the following: art (Erasmus, 2012; Wu, Chen, & Huan, 2010); food (Dimitrova & Yoveva, 2014; Blichfeldt & Halkier, 2014); fashion (Weller, 2013); film (Meias et al., 2011; Park, Lee, & Park, 2011); flowers (Deng & Pierskalla, 2011; Tkaczynski, 2013a); wine (Marais & Saayman, 2011; Novelli, 2004); animation (Ruling, 2011); jazz (Curtis, 2010); homelessness (Rushford, 2013); winter (Gabbert, 2011; Wardrop & Robertson, 2004); spring (Li, 2014); music bands (Gouzouasis & Henderson, 2012); religion (Calestani, 2013; Roemer, 2007); and music (Anderton, 2011). Further, festivals can be combined, as in the case of religious and music festivals, expanding their potential attraction and goals (Tkaczynski & Rundle-Thiele, 2013).

Festivals can physically alter an urban landscape, encouraging the use of spaces for post-celebratory events (Derrett, 2004), facilitating a capacity-enhancement approach. Festivals also are increasingly being used by community organizations to strengthen local social capital and the transmission of intergenerational knowledge, civic pride, and well-being, as in the case of Indigenous festivals in Australia (Slater, 2011). Festivals can span lengthy periods of time, in some cases up to two weeks

(Grunwell, Ha, & Swanger, 2011), reaching a broader audience than parades or fairs.

Festivals are usually of short time duration (less than two weeks), involve sizable crowds, are held during a particular period during the year, and are limited to specific areas of a community (Kim et al., 2010), although geography can be considerably larger in area than for fairs. These qualities make festivals attractive to sponsor. Community groups must think strategically about staging festivals over other forms of events, and where they should be held. Washington's La Fiesta DC, for example, is held in the neighborhood rather than in a prominent place such as the National Mall, a decision dictated by festival goals. However, the National Mall in Washington, DC, is a cohesively built environment and conducive to events and reflexivity (Evans & York, 2013; Lucas, 2014).

It is important to understand where festivals are usually held. The possibilities seem endless, based on the study by the US National Endowment for the Arts (2010), which illustrates the venue flexibility associated with festivals. The communities studied by the NEA used the following venues in staging a festival: plazas or parks (46% of the communities), streets (25%), outdoor theaters (19%), other (18%), museums (10%), concert halls (10%), schools or colleges (9%), public buildings (8%), private grounds (8%), community centers (7%), waterfronts (6%), restaurants and hotels (4%), fairgrounds (4%), churches (3%), and sports venues (2%). Public places such as plazas, parks, and streets were the locations that accounted for almost three-quarters of the festivals (71%) held and highlight how public places can fulfill multiple roles in urban communities.

Festival locations can cover a variety of urban geographical settings. Children's playgrounds have been found to be favorable sites for celebrations, particularly fairs and festivals, although they are not explicitly included on the list above (Elsayad, 2014; Lim & Barton, 2010). Festivals held near bodies of water (see waterfronts as the location for festivals in 6% of communities, above) bring an added dimension to celebratory spatial properties and can introduce water activities as an enhancement to the types of activities typically found for land-based festivals (Volker & Kistemann, 2013).

Rewards and Benefits of Festivals

Festivals are widely viewed as representing social and cultural phenomena. Known as the "living culture" of a community, festivals provide a window through which the external community can view and understand the festival community in a manner that has historically eluded them. Festivals are vehicles through which "the history of a community can be narrated and a sense of place uniqueness can be established" (Lau & Li, 2015). Creating and maintaining civic pride can be a goal of festivals (Whalen, 2010), resulting in community building (Derrett, 2003, 2004). Often festivals are thought of as being beneficial to local economies (Kostopoulou, Vagionis, & Kourkouridis, 2013).

Festivals that result in urban renaissance or renewal require cultural capital as a foundation for place-based promotion and events (Betancur, 2011; Jamieson, 2013). Broadening cultural capital facilitates an understanding of how it can manifest itself in both tangible and intangible forms (del Barrio, Devesa, & Herrero, 2012, 235). Unfortunately, this also increases the challenges for how this capital can be evaluated. A focus on cultural capital requires a nuanced understanding of local institutions and their fostering of this capital through the sponsorships of events (Khan, 2013).

Festivals can be sources of innovation and creativity, including community and institutional renewal (Larson, 2011), bringing rhythms (interactions) that transform urban spaces (Duffy et al., 2011). Festivals, more so than fairs, can evolve from local to national and even international events (Ferdinand & Williams, 2013), highlighting their potential to cross geographical boundaries. This evaluation brings its share of tensions and loss of local control over the festival.

Derrett (2004) identifies an impressive list of the benefits that festivals can bring to a hosting community: festivals can be a vehicle for local pride, urban renewal, conservation, civic design, generation of employment, economic development, and identity. Festivals are attractive because of the range of local intra- and inter-community goals they can address (Derrett, 2004), particularly when combined with street parades and other forms of celebrations (Arroyo-Martinez, 2012; Batty, Desyllas, & Duxbury, 2003a). Festivals can enhance perceptions of local identities (reputation), although communities must not have overly high expectations because events are not a panacea (Croy, 2010; De Bres & Davis, 2001; Jaguaribe, 2013b).

Festivals can be sources of economic capital, necessitating investments, resulting in competitions to stage them because of their importance (Van Aalst & van Melik, 2012). Lukić Krstanović (2011) introduces the role and function of political folklore in festival events, adding an often overlooked dimension.

Festivals, regardless of theme and size, benefit communities, particularly in the case of marginalized communities, by enhancing their health, education, and employment opportunities (human capital) and providing residents with an avenue for meaningful participation (Phipps & Slater, 2010). The economic benefits of festivals, however, predominate in the literature (Gibson et al., 2010).

Moscardo (2007) broadens these benefits, based on an analysis of the role of festivals in regional development, but these are also applicable for urban areas. Moscardo identifies the following three critical constructs that play a role in achieving this goal: building social capital, enhancing community capacity, and providing support for products and services not related to tourism. These constructs fit within a capacity-enhancement paradigm by stressing local assets or capital.

Getz (2009) argues that the values or capital that festivals bring to communities warrant greater public policy attention and support. Community stakeholder participation in festivals can translate into other ventures and collaborative partnerships, benefiting other underdeveloped sectors of a community (Bishop, 2012; Robertson, Rogers, & Leask, 2009).

The definition of *stakeholder* must go beyond the conventional uses of the term to include individuals who play leadership or opinion-setting roles although they are not elected officials or heads of major community organizations; and it must be kept in mind that stakeholders may have multiple roles and conflicting agendas (Reid, 2011).

Schools can collaborate with community organizations to organize festivals involving students and teachers (Gandy, Pierce, & Smith, 2009). Festivals can be a social capital bridge between students, families, communities, and schools (Kroeger, 2014). Classroom celebrations illustrate how these events can vary in size, goals, and with respect to the ages of students (Isoldi et al., 2012). Festivals can, however, present opportunities to exclude ethnic and racial groups (Clarke & Jepson, 2011), raising the importance of a subtle understanding and perspective on these events, and how local circumstances shape these celebrations.

University and college students can be a vast and often overlooked resource in working on community festivals. The University of Cambridge issued a report entitled *University Engagement in Festivals: Tips and Case Studies* outlining the potential contributions students can make to communities (Buckley, McPhee, & Jensen, 2011, 3): "Festivals are a growing part of the UK's cultural life ... As a result, festivals can offer a valuable opportunity for students and higher education institutions to engage the public with their work, and to promote the activity and benefits of higher education. They enable students and staff to encounter members of the public face to face and make use of diverse formats for public engagement." Civic engagement and learning go hand in hand, providing students with opportunities to practice within community settings without suffering the stigma and suspicion often found when universities enter communities under the guise of helping (charity paradigm).

Institutions of higher learning often have service-learning departments or service-learning projects focused on helping communities with fairs and other celebratory projects (Delgado, 2016). Graphic art and art departments can assist with supporting artwork, and media departments can help with media relations and press releases, for example. Youth bring vision, energy, and creativity – all qualities that enhance celebrations.

Highly marginalized communities seem to be in a perpetual state of searching for potential resources, and universities are often at a loss as to how to engage communities in an affirming manner that allows them to carry out their missions. Having students engage in festivals represents a win-win for communities. These collaborations open up doors for potential community residents to see higher education as a viable option rather than a foreign experience (Carney et al., 2011). These relationships can assist academics in establishing relationships for conducting research and attracting external research and funds for community investments.

There are rewards for volunteering at community festivals (Barron & Rihova, 2011). Volunteers may be motivated by being part of the community, enhancing their competencies, having access to attractions and behind-the-scenes activities, being invited to parties celebrating successful events, and even gaining college credit for engaging in community service, as well as getting social recognition (Wen, Yang, & Huang, 2013). Festival attendance motivation has been extensively research. Motivation for volunteering has not, and the findings can prove beneficial for a community over the long run by increasing in prominence as civic engagement plays a central role in the staging of festivals.

Tensions and Challenges of Festivals

There is an increased understanding of what factors are likely to lead to the success of a community festival (Lade & Jackson, 2004). Adequate funding is often a challenge in making festivals sustainable (Lee & Kyle, 2014). Sustainability does not occur by chance. Contrary to popular beliefs, failures of community festivals are quite common (Getz, 2002). Festivals often fail because of limited budgets and poor marketing strategies (T.H. Lee, 2013). A lack of clarity on primary goals, too, compromises the chances of success.

Festivals can release pent-up tensions resulting in "battlefields of contentions" (Gotham, 2005). These fractures take on public prominence when they get manifested through this form of public event because the eyes of the hosting and broader community are focused on these celebrations (Finkel, 2009).

Because the size and complexity of this celebration does not have the flexibility of the staging of fairs, festivals can be considerably more challenging to stage. Unlike for fairs and parades, organizers may have to charge entrance fees for festivals to avoid overcrowding and off-set the costs of certain attractions with the potential to attract large crowds and publicity (Abu, 2012).

Festivals may have a central goal to generate funds for a worthy cause, or of bringing added publicity to a cause. This necessitates having staff with competencies generally not found among community residents, for example, having someone with expertise in publicity. Expertise in the pricing of activities, goods, admissions, and in determining how payments (credit cards, debit cards, cash, and checks) will be processed dictate a set of competencies that may be missing within a community of volunteers. Festivals can involve the selling of alcohol and thus result in alcohol-related acting out (Németh et al., 2011).

Large festivals, particularly those with worldwide followings and reputations, must not be viewed as a single event. Instead, as with Carnival in Rio De Janeiro, Brazil, festivals may appear to be a singular event to the outside world but that is far from the case if viewed locally, where individual events taking place around the city have unique characteristics (Sussman & Barnes, 2014). Large festivals provide opportunities for stakeholders to assume greater prominence due to the social and political complexities inherent with large-scale celebratory events (Getz & Andersson, 2010; Köster, 2014).

National-level large-scale events, as in the case of North Korea's Grand Mass Gymnastics and Artistic Performance Arirang and Gabon's Independence Jubilee, strive to create alternative narratives of a country before a world stage (Fricke, 2013; Merkel, 2013), highlighting the public relations aspect of events. "Big stake events" can draw major stakeholders, relegating local stakeholders to a secondary status (Reid, 2011; Ziakas & Costa, 2011).

Festivals represent big gambles and bring potential benefits and risks (Grames & Vitcenda, 2012). For example, which groups are attending festivals and how much they spend while there has started to get increased attention (Taks et al., 2013). Festivals can be staged in response to efforts to achieve inclusion and respond to changes, helping to ensure continued relevance and meaning to changing communities and times (W. Lee, 2013). Waitt (2008) raises caution about urban fairs not being a panacea for the social and economic ills of cities. Governmental efforts at espousing the virtues of private-public partnerships to boost city images of urban festival spaces and liberating certain social groups can also result in constraining and further oppressing other groups.

Festivals can boost local economies or help drain them; they can foster community pride or shame, and can strengthen or weaken social and political relationships. Festivals can generate learning or be counterproductive to collective learning. The potential of big rewards is not possible without the potential for big disappointments. Festivals can provide opportunities for local employment, which only increases in significance in highly economically distressed urban areas (Litvin, 2013).

Stadler, Fullagar, and Reid (2014) address the emerging concept of "festival knowledge" to capture the growing complexity of festival management and the increasing pressure placed on organizational sponsors. However, with increased pressure on knowledge acquisition, there will be increased benefits to the field.

Evaluation and Research Challenges of Festivals

Evaluators' lives would be simple if their only festival evaluation task was to focus on the number of tickets sold (Saayman, Kruger, & Erasmus, 2012). However, as evidenced by how festivals are conceptualized in this book, ticket sales are but one facet and not even a major facet of evaluation.

There are many different facets related to evaluating festivals. The management of these events has been signaled out as being of great importance (Getz & Frisby, 1988; Schulenkorf, 2012), as has the level of support for the community (Phipps & Slater, 2010), quality of experience (Tkaczynski, 2013a, 2013b, 2013c), level of meaningful collaboration (Cabral, Krane, & Dantas, 2011; Ziakas & Costa, 2010b), emotional and functional values (Lee, Lea, & Choi, 2010), and local economic benefits (Okech, 2011); there are also many intangible or invisible factors (Araoz, 2011; Reverte & Izard, 2011; Ziakas, 2010) such as festival meanings to residents from a variety of viewpoints including cultural heritage (Ziakas & Boukas, 2013) and place attachment.

Tkaczynski (2013c) addresses stakeholders influencing who attends festivals and why this is essential to understand who actually attends and how they are influenced, with important implications for tourism. Evaluators must broaden their scope on attendees' motivation, including how stakeholders influence them (Buch, Milne, & Dickson, 2011).

Derrett (2004) reintroduces the concept of space and its importance regarding festivals, which evaluators must not overlook in developing a comprehensive understanding. Once a festival is over, communities may be inspired to continue using the space where the festival transpired for other purposes. It is insufficient to focus evaluation on the activity itself during the time span in which it unfolds. Follow-up is essential to capture the long-lasting benefits of a festival, and that includes how the space where it was staged was enhanced or transformed.

There will be resistance to evaluating events because of the resources that must be devoted to this activity, as well as concerns about what the evaluation might reveal, which can compromise the goals that event organizers sought to accomplish, and their reputations, too (Swan & Atkinson, 2012). The role of "politics" is very much integral to any evaluation process, and celebratory events are not exempt.

Festival creation and attendance takes on significance when one of the goals is to make them community-centered. Rogers and Anastasiadou (2011) propose the following five-point framework for measuring festival experience and community involvement that introduces institutional engagement as a critical element: involvement of schools, volunteering (civic engagement) opportunities, participation in decision making, accessibility, and business cooperation.

Having community institutions alongside resident engagement expands the concept of participation, raising the importance of key institutions, which can be formal or informal. Richardson (2011) places importance on food at school celebrations as a means of fostering school-community partnerships, civic engagement, and increasing the understanding of cultural capital.

Boo, Carruthers, and Busser (2014, 269) take a novel research approach toward festival nonparticipants and participation constraints (structural, interpersonal and intrapersonal) by illustrating how different types of constraints result in different types of outcomes in festival attendance. These authors broaden the scope and discussion of constraints and negotiation to take into account the context of the festival and differences between interested and disinterested event nonparticipants to increase participation and the benefits of festival attendance.

Gardi (2014) discusses the importance of capturing festival visitor satisfaction and the need for a practical and user-friendly method: "It is crucial for festival management to monitor and evaluate visitor satisfaction in order to understand and identify the needs and perceptions of attendees, which in turn allows organizers to design and tailor the festival elements towards them, leading to higher visitor satisfaction, positive word-of-mouth advertising, and increased likelihood of repeat attendance."

Understanding participant motivation translates into a targeted marketing campaign (Luchian, 2014), helping organizers to grasp future potential attendees' behavior. Manthiou and colleagues (2014) identify four underlying dimensions of festival attendees' experience – education, entertainment, esthetics, and escapism – as having a positive effect on vivid memory, a key component of loyalty. The manifestation across community subgroups, too, must be captured to develop a comprehensive understanding of the importance of festivals in a community's life.

Community Practice Implications and Festivals

Festivals bring potential for community practice undertaking large-scale and highly public events to showcase community solidarity and assets. The event scale ensures that there will be wide-scale publicity, allowing the development of goals that are closely tied to public image and messages. Festivals are more than a publicity opportunity, however, bringing the potential of accomplishing a range of goals, including enhancing community assets.

Émile Durkheim made early reference to festivals, in *The Elementary Forms of the Religious Life* (1915), as vehicles for intensifying the collective being, signaling sociology's potential contribution to this subject, although this discipline has yet to realize its potential in this regard (Sassateli, 2011). Festivals have benefited from considerable attention from scholars in other disciplines, although much work remains to be undertaken to provide a worldwide inventory (McDowall, 2011).

Film festivals, for example, have a long tradition worldwide with some having an international reputation (Park, Lee, & Park, 2011; de Valck & Soeteman, 2010). These festivals may have started as community ventures, but many were quickly usurped by local governments and transformed into major tourism events. Further, film festivals rarely target marginalized communities, and they may have limited appeal in urban communities with high concentrations of newcomers, for example, unless they target them specifically.

Festival literature presents conceptual and methodological challenges in creating a consensus understanding of their universal qualities, and how local sociocultural circumstances shape goals (Slater, 2010). It becomes a challenge drawing together different disciplines for developing a common understanding for the furthering of social utility and capacity enhancement (Chapple & Jackson, 2010; Ziakas, 2013a, 2013b).

Complicating arriving at a consensus view of festivals further is that there is no prevailing way of looking at festivals that enjoy universal acceptance. Williams and Bowdin (2010) found that there is no one model for festivals. The flexibility in how communities think about festivals, including their longevity and taking their socioecological circumstances into account, makes classification challenging but essential to increase our understanding of the functions they play (Gabbert, 2011; Mangia et al., 2011; Slabbert & Saayman, 2011). Urban festivals are increasingly being used for creating images and reshaping urban social processes (Johansson & Kociatkiewicz, 2011).

De Bres and Davis (2001) argue that small festivals challenge common perceptions of local identity, with attendance being highest at the opening and closing of the festivals, requiring greater attention to how to maintain a high level of attendance throughout the life of an event. Small community festivals may be more frequent, representing the base of a pyramid, and very few, more infrequent, but considerably larger festivals represent the apex of the pyramid (Andersson, Getz, & Mykletun, 2013a).

Gibson (2014) concludes that small music festivals have the potential to wield prodigious impact. The larger the festival, the longer the history when compared with the small and more frequently held festivals. This observation is applicable to fairs and parades as we discuss evolution and sustainability.

Olsen (2013) argues that urban festivals are places where communities can experiment with innovation and explore alternative cultural strategies. These experimentations can alter local community identities and facilitate the integration of the latest trends (Newman, Dale, & Ling, 2011). For example, efforts are underway to make more festivals green and sustainable, introducing slow food (healthier eating), and tying these events to broader environmental moments (Barber, Kim, & Barth, 2014; Ensor, Robertson, & Ali-Knight, 2011; Frost & Laing, 2013; Laing & Frost, 2010; Okech, 2011).

Highlighting how festivals support the greening movement represents an attempt at drawing audiences who share these concerns (Horng et al., 2014; Laing & Frost, 2010), although formidable barriers to achieving this goal remain (Ball & Wanitshka, 2014; Mair & Laing, 2012). Festivals can be spaces where unexpected and new information can be obtained while people are having fun (Schulte-Römer, 2013).

Understanding the origins and histories of festivals provides a sociocultural contextual window into how, and why, communities put emphasis on an event, even though they can be challenging from a political and logistical perspective (Calestani, 2013; Carlsen, Robertson, & Ali-Knight, 2007; Pipes, 2013; Slabbert & Saayman, 2011). Chacko and Schaffer (1993) examine the origins of the New Orleans Creole Christmas festival, which lasts one month, and find that it was the intent of the tourism industry to attract visitors during the slowest month for tourism. Although this festival accomplished the economic goals that led to its creation, any social and cultural goals were unrealized.

Quinn (2006), too, raises concerns about how economic goals subvert social and cultural goals, dramatically altering the original purposes for initiating festivals. Economic goals are rarely center stage early in the creation of a festival, although these can never be totally ruled out as a key goal. It is advantageous to frame an issue in terms of economic benefits to obtain political support, particularly from a legislative body or from elected officials, and that often happens in the case of festivals (Harnik & Crompton, 2014).

Complex festival goals provide fertile grounds for breeding discontentment or an opportunity for tensions to surface because the more goals, the higher the likelihood of having different interpretations. Crespi-Vallbona and Richards (2007) address this point and discuss how cultural festival tensions can be divided between stakeholders with public policy interests favoring economic and political goals, and cultural producer stakeholders who emphasize social aspects related to cultural identity. Festivals can address these two schools, but cannot concurrently emphasize both without causing serious tensions.

Community events can be a draw for anti-social behavior (Deery & Jago, 2010), an aspect of celebrations that few communities wish to foster. Tensions, too, can be found between groups favoring putting forth an acculturated identity and values versus those who favor traditional identity and values. What may appear to be a simple difference of opinion can, in fact, be quite disturbing and lead to generational conflicts. Parading can exacerbate community tensions rather than relieve them

(Barrios, 2010; Bell, Radford, & Young, 2013), and do so in a highly public manner, further exacerbating local circumstances.

Ethnic or nationality festivals are not monolithic, even when sponsored by the same group, and this is well illustrated by Chicago's Turkish community. Girit Heck's (2011) study of two Chicago Turkish festivals (the Chicago Turkish Festival and the Chicago Turkish World Festival), show how on the surface festivals may appear similar even though they, in fact, represent different political and commercial goals. The political goals of these two festivals were to establish and maintain positive relationships with state officials and be a lobbying mechanism for Chicago's Turkish community. Their commercial goals were different and multifaceted, with the one festival seeking to generate new streams of income for local Chicago Turkish artists and entrepreneurs, while the other festival was attending to transnational businesses from Turkey and other countries. The glorification of Turkey's Ottoman history wields great influence in one festival, while the other emphasizes Islamic influences on culture and national identity.

Rosendahl (2012) addresses another dimension of festivals, in discussing pride festivals, which emphasize groups that have been marginalized, and how the festivals are meant to affirm and announce their political arrival on the public scene; however, he observes that while the festivals can create a unified group identity, they can also become a space where certain identities within the marginalized group are prioritized, at the expense of gendered and racialized groups. Creation of group unity can be found in various pride parades. The 2014 New York City Pride Parade marked a historical moment when the Boy Scouts of America in full uniform wearing rainbow neckerchiefs marched, making it the first time that the Boy Scouts took part in this parade, following a vote taken by the Boy Scouts to allow gay scouts to openly participate. Incidentally, the parade passed the historic Stonewall Inn, the bar widely credited as being the birthplace of the gay rights movement in the United States (Yee, 2014). This social justice symbolism was not lost on the parade marchers and bystanders.

The transition of LGBTQ groups from enclaves and isolation to being a part of a broader community can also be manifested through sponsorship and active participation in pride events (Doan & Higgins, 2011; Kuhar & Svab, 2014; Menon, 2013; Nash & Gorman-Murray, 2014), including the seeking of political rights (Ferry, 2012). Pride events provide gay couples an opportunity to bring their children to have an affirming experience pertaining to the sexual identity of their parents

(Rothblum, 2014), which can be part of a broader campaign of social capital bonding and bridging.

These events solidify community identities and generate social power (Bruce, 2013), and are an opportunity to garner support from sympathetic majority group members (Ratcliff, Miller, & Krolikowski, 2013). However, attendees at pride parades can still face insults or ridicule, compromises to safety, and political backlash as in the case of the 2007 event in Santiago, Chile (Barrientos et al., 2010) and in Southeastern Europe (Pearce & Cooper, 2014), as well as at the 2013 Pride Parade in Belgrade, Serbia (Dzombic, 2014; Embassy of the United States Serbia, 2013).

A philosophy of inclusion can nevertheless bring tensions and further marginalization (Chandler, 2013; Fujimoto et al., 2014; Kim, 2013). When exclusion occurs, an event's sustainability (cultural, social, and environmental realms) can be compromised, creating hurt feelings and negative collective memory on the part of excluded groups (Flecha et al., 2010; Myers, Budruk, & Andereck, 2010; Jepson, Wiltshier, & Clarke, 2008). Community support and attachment can be expected to play influential roles in achieving sustainability (Duran, Hamarat, & Özkul, 2014; T.H. Lee, 2013).

Conclusion

Festival image will always be synonymous with celebratory events, making a festival a powerful event for a community to sponsor, participate in, and watch. Festivals provide opportunities to be inclusive under the best of intentions, or exclude groups under the worst of intentions. The social and political manifestations of festivals span national boundaries, with implications for diaspora communities hosting an event and for communities back home for those who left their original homes for new ones. The sheer magnitude of major urban festivals brings challenges in planning, funding, managing, and evaluating these events.

Festivals, too, can include groups sharing varied social justice agendas. This does not mean that these efforts will not engender political backlash since their high profile makes them attractive for demonstrations. Nevertheless, there is no denying the rich and illustrative history of communities using festivals to publicize their progress.

7 Parades

Parades stop traffic and do so with flair, while reaching millions of people. Parades lack the intimacy of fairs and even festivals because of how they are staged and because they unfold over a large tract of geography. Nevertheless, parades have the potential to span broad social, political, and cultural issues, and imprint their importance for sponsoring urban communities and on the broader society that attends or views them on television or via some other media outlet.

No celebratory event takes in such a vast public space, necessitating an examination of parades from a vantage point that differs from fairs and festivals. High spectacle parades are also high stakes gambles made by communities to advance their primary goals, and may result in high gains for the communities and groups sponsoring them, for example, in the case of heritage-themed parades. Parades entail complex logistical planning and bring high costs, both financial and psychological. This translates into high excitement or high anxiety for the sponsoring organizations. The key personnel in charge of coordinating parades take on even greater significance, and much will be demanded of them in staging a parade than for a fair or a festival.

Definition of Parade

Definitions of parade bring elements that can be found in definitions of fair and festival, yet the uniqueness of a parade sets it apart. Parades can be defined as "planned rituals that allow people to display collective identities publically ... break[ing] the rhythm of everyday life and giv[ing] collective expression to people's joy, sorrow, hope, claims, or aspirations" (Loukaitou-Sideris & Ehrenfeucht, 2009, 61). This simple definition hides its complexity because of the role that sociocultural factors play in the origins of parades (Newman, 1997; Olsen, 2013). Almost all countries and geographical entities have parades to mark a significant date in their history, highlighting how parades can play a role in galvanizing social and political forces.

Parameters and Range of Types of Parades

It may appear that parades share so many similarities that it would be arduous, if not fruitless, to differentiate between various types other than the most obvious elements related to length, attendance, and media coverage. However, it would be a serious mistake to take this perspective. The social, political, and cultural themes of parades sometimes are quite obvious and at other times well disguised, necessitating an in-depth understanding of how historical and local context have shaped their evolution and current manifestations, including sustainability.

Parades may seem predictable and very similar to each other to the untrained eye. They all seem to have marching music, participants in costumes, streets cordoned off, large banners, hawkers, and local celebrities and elected officials at the front of the parade. However, parades are certainly not homogeneous events, and they represent significant challenges in analyzing them and their impact on communities. Military parades, for example, bring a different feeling and experience for attendees when compared to nationalistic parades with an ethnic or racial theme.

Manhattan, New York City, hosts approximately 180 annual ethnic parades (Malek, 2011), which translates into just over three parades per week year-long. However, the parades are concentrated over three seasons (excluding winter), creating a ubiquitous presence of parades over a densely packed area of the city. Malek (2011), in addressing New York City's Persian Parade, refers to such parades as "public diaspora parades," and submits that these celebrations present a picture of a group to the outside world, as well as being a vehicle for helping multiple generations negotiate their identities. Parades, however, also can be viewed from a public demonstrations/activists perspective when they specifically embrace a social justice agenda by redressing an injustice perpetrated by the broader community on the community sponsoring the parade. For example, Manhattan has also been the center of a bicycle protest parade (Blickstein, 2010). Parades as a form of public demonstration are also a form of theatrical performance, with elements related to spectacles taking on significance (Rosental, 2013).

132 Community Celebratory Events

Rewards and Benefits of Parades

A public community-sponsored spectacle that can stretch out for a large number of city blocks, and even miles, with requisite media coverage, provides an ideal circumstance for projecting a well-defined image to the broader community, and serves as a rallying point for a community to come together. The rewards and benefits inherent with parades can be significant for stakeholders and residents alike, particularly since the entire effort lasts a relatively short period of time that can be measured in hours, but can require a year of planning.

Sponsoring a parade can be considered a public declaration that a group or community has achieved a high level of power. The planning and launching of a parade reflects complexities and organization that can best be accomplished by a community that has come together in pursuit of a common agenda, enabling participants to become part of the community (Huang, Li, & Cai, 2010). This declaration of status may open old wounds or create new ones.

Ethnic parades can bring many different groups together. But there can also be fragmentation or, for example, situations where Latino groups with different countries of origin sponsor their own parades in addition to participating in a pan-Latino event, as happens in Chicago. During a three-week summer period, Chicago stages three Latino festivals: Viva Chicago, Fiesta Boricua (Puerto Rican), and the celebration of Mexican Independence. Each of these festivals shares similar goals but also differences based on the sociopolitical backgrounds of the sponsoring communities (García & Cantú, 2007).

Sengupta (2012) notes that the "American ethnic parade" has multiple meanings and layers and is an activity that can generate pride, identity, emotions, and reveal hidden sides to a community. Further, parades disrupt daily routines in the lives of participants and spectators, creating an opportunity to learn and share with those outside of the ethnic group sponsoring the event. Ethnic parades in particular but also other types are a mirror on a community, but like mirrors that purposefully distort images, parades can project images internally and externally that are both intended and unintended.

The grandeur of a parade is inescapable, and its potential for covering a large geographical distance of multiple miles, traffic intersections, and neighborhoods is part of its appeal, and the visual spectacle makes parades particularly appealing for television coverage. The world seems to stand still when there is a parade, and there is no disputing how it disrupts daily events and moments for those who are spectators or whose daily routine is interrupted because of a parade route. Consequently, it is easy to identify numerous rewards for staging a parade.

Community spirit can be manifested through an event, even against environmental and circumstantial obstacles. Grams (2013, 501) undertook a study of New Orleans parading and the role that parades have played there in the period after Hurricane Katrina. These parades were successful without depending on a functional city structure for support, and they created a "logic and momentum for rebuilding communities and meaning in local life." Two parades stand out (Mardi Gras Indian Tribes and Social Aid and Pleasure Clubs) for providing a marker of the city's recovery and for maintaining cultural consciousness of self and New Orleans.

Few parades are as ubiquitous on the East Coast of the United States as St Patrick's Day parades, with the ones in Boston and New York City standing out. However, the social contexts in which these parades unfold are dramatically different reflecting a host of sociohistorical factors. Boston's St Patrick's Day Parade takes place in South Boston, a community with a long history of Irish roots. Controversy has followed this parade since its organizers refused official entries from LGBTQ groups. New York City's St Patrick's Day Parade takes place on Manhattan's Fifth Avenue, easily accessible to the city's population (Ridge & Bushnell, 2011). In essence, in New York, everyone is Irish on St Patrick's Day! Marston's (1989) social historical study of St Patrick's Day parades in Lowell, Massachusetts, from 1841 to 1874 found that in that period these parades were best understood as a public expression of community power. Initially, they were meant to impress the Yankees and Irish, and eventually they evolved into an opportunity for greater political participation (political capital).

Finally, Williams (2012) studied Chicago's annual South Side Bud Billiken Day Parade, established in 1929, which has since grown to have over one million spectators and a televised audience of 25 million; it is considered to be the largest Black parade in the United States. This parade encompasses the usual goals of nationalistic parades. Efforts to make it more inclusive of alternative Black family definitions met with resistance, however, as when in 1993 a group of Black lesbians decided to join the parade in an open manner; their participation was ultimately achieved through the courts.

Parades serve as a very public focal point for communities, bringing sought attention or publicity, because this attention is often a central goal, but their symbolic meaning can be expected to differ by neighborhood (Loukaitou-Sideris & Ehrenfeucht, 2009). Contextualization allows for a deeper appreciation and understanding of why parades are important to the communities sponsoring them.

Parades publicly affirm how far a community has progressed and why the immediate world needs to take notice (a form of political capital) and provide community residents with endless possibilities for tapping their cultural assets. Community artists (cultural and economic capital), particularly those in highly marginalized communities, have few outlets for their talents in the broader community. Limited outlets may be the result of limited social contacts. Parades, however, provide a venue for showing the world the community's talents (Markusen, 2007).

The process of constructing and decorating floats, for example, is a way of tapping local artistic talent and, in the process sustaining community bonds between generations and defining local identities (Kuroishi, 2013). These floats require sponsorships and are labor intensive in their construction. Parades provide local artists with an opportunity to make art for public purposes, exposing a different side to their friends and neighbors, and reflecting community cultural assets (Hoefferle, 2012).

Tensions and Challenges of Parades

Discussion of tensions and challenges in staging parades will invariably focus on the political challenges of bringing large groups of people together to plan, stage, cover, and attend a parade. The logistics associated with parades, too, stand out in importance. Any reader who has planned an event, be it a surprise party, wedding, family reunion, or conference, for example, will attest to the sleepless nights that occupy such planning. Personalities, time, weather, competing events and interests, and securing permits, venues, and food will all elicit compassion from those who listen in a sharing of horror stories about their own experiences.

Spectacular parades bring with them spectacular tensions and challenges. The grandeur and the disruption of daily routine brings with it tensions and challenges for sponsors, even when a parade is carried out flawlessly. When tensions that exist within the sponsoring community erupt on a public stage, parades have the potential to draw unfavorable light on the community by exposing its "dirty laundry" to the general public. How these events are covered by the media, particularly in highly politicized environments, reflects biases and feeds into making these events even more politically charged "high stakes" ventures (Duguay, 2013; Ferman, 2013).

A historical review of community parades reflects an evolution that also highlights tensions. The evolution of community parades can result in empowering disenfranchised groups to view these events as a vehicle for pursuing social justice. Community events and ethnic establishments, for example, may give the outsiders the impression that the community is united socially and politically. These events can be viewed as manifestations of how the community and country is being transformed in a highly positive manner. Nevertheless, a nuanced examination may reveal deep intragroup tensions with extensive histories dating back to a community's origins (Cunningham & Gregory, 2014; Portes, 2012).

Large-scale events such as parades can provide a highly public stage for violence or potentially tragic events, as in the case of the 24 July 2010 Love Parade in Duisburg, Germany (Helbing & Mukerji, 2012; Zhang et al., 2013). Concepts such as "intentional pushing," "mass panic," "stampede," and "crowd crushes" have emerged to develop a better understanding of how tragedies related to crowds emerge in community celebratory events such as parades (Fradi & Dugelay, 2013; Pretorius, Gwynne, & Galea, 2015; Schwarz, 2012).

The following account of the Love Parade emergencies illustrates the potential scale of crowd-related tragedies (Ackermann et al., 2011, 484):

> An estimated 250 000 people took part in the Love Parade 2010. According to the emergency services' plans there were 30 first-aid posts on site, each with capacity for 10 patients, as well as one on-duty physician, 20 first-aiders, and one ambulance, and in addition, two standard emergency treatment stations for 50 patients each (BHP 50, a category established by the German Institute for Standardization [DIN, Deutsches Institut für Normung]). These facilities registered a total of 5600 patient contacts over 24 hours. The 1600 members of auxiliary staff were increased to approximately 4000 after news of the mass panic was received. Seventy ambulances and other emergency vehicles and nine emergency helicopters were used.

Cao (2011) estimates that approximately 1,500 incidents of violence occur during parades each year. Even the possibility of falling bullets fired during celebratory events can result in casualties and injuries (Al-Tarshihi & Al-Basheer, 2014; Rapkiewicz, Shuman, & Hutchins, 2014). Celebratory parade violence (CPV) attributes such violence to either deindividuation (euphoric and mindless state) or impression management (reduced accountability/self-awareness), raising implications for how parades are managed and policed. The potential of dangerous conflict is ever present when events involve large crowds.

Parades can expose community fractures and serve as a focal point for drawing criticism, as evidenced in New York and Boston with their St Patrick's Day parades and historical exclusion of formal LGBTQ participation. Media coverage about this resulted in major corporate sponsors withdrawing their support until this group is officially included in the parades. The following statement issued by the Boston Beer Company, a former corporate sponsor of Boston's St Patrick's Day Parade, highlights this key point (quoted in R. Murray, 2014):

> As a local business, supporting our Boston community is very important to us. The St Patrick's Day Parade is just one of the hundreds of events and organizations we support in and around Boston. We provide charitable donations to organizations in South Boston and around the city that address critical needs including supporting the arts, the environment, building communities, veterans initiatives, youth leadership development and addressing educational disparities, just to name a few. We also provide support for a number of organizations whose primary focus is supporting civil rights, the LGBT community, marriage equality, and the Boston Pride Parade. We have deep roots in Boston and will continue to support local charities here and across the commonwealth, especially those charities that our employees and drinkers find meaningful and impactful. That being said, our namesake, Samuel Adams, was a staunch defender of free speech and we support that ideal, so we take feedback very seriously. The majority of our commitments are year-to-year, and we will continue to evaluate each organization and event before making additional contributions.

The statement illustrates how social factors can result in the re-examination of corporate sponsorship of public celebratory events, and the precarious nature of this relationship for event sponsors as they try to navigate socially controversial topics.

Tensions and challenges are an inherent aspect of celebratory events, and they may simply be impossible to eschew. That, however, does not mean that they cannot be minimized through careful planning. Parades have high potential for magnifying tensions and tragedies, and event planners cannot rectify situations the next day after they go dramatically wrong. They must wait another year to do so. Further, the presence of local media will capture miscues and misfortunes, further compounding efforts to control the message that emanates from events gone wrong.

Evaluation and Research Challenges of Parades

The evaluation of the impact of a parade brings additional challenges for evaluators because of the wide geographical area involved, since these events are not concentrated in a narrow area, and they lend themselves to participation across a wide spectrum of people from causal onlookers to active participants. Parades tend to be high-profile events that draw considerable media coverage, which can complicate the assessment of their impact on community and on external viewers, but also offer a source of data for analysis, although assessment of news coverage is a skill-set that few evaluators possess.

Using participant observations in evaluation broadens the possibilities of involving community residents in an active and meaningful manner, facilitating the targeting of different subgroup perspectives (Mackellar, 2013a, 2013b). These qualitative insights answer questions that can only be answered through this form of data gathering, as is the case with photovoice and the narratives that accompany photographs (Delgado, 2015).

What makes parades such spectacles and so appealing also translates into evaluation challenges in determining their success beyond noting crowd estimates, number of floats, how far they reach, and number of participants. A "bean counting" mentality masks a parade's socio-economic and sociopolitical reach, and thus its potential impact on a community. Economic impact must be measured in both the formal and informal economy to have a comprehensive understanding of this impact. The extent and nature of media coverage brings a perspective that goes beyond what the parade was about, to a content analysis of this coverage.

Community Practice Implications of Parades

There are few childhood events that can compare to the experiences of attending a major urban parade. The anticipation of attending meant that it was difficult to get a good sleep the night before. These memories of parades are a part of many childhoods. California's Pasadena Rose Day Parade is a nationally televised event, as is Macy's Thanksgiving Parade (Grippo & Hoskins, 2004), and I can remember attending more than my share of these when I was a child in New York City. The floats were larger than life, and the balloons featuring well-known animation characters filled the skies; the pageantry was simply indescribable, with all spectators, regardless of their age, being entranced.

Parades enjoy an international following for good reasons, and they are often used as a principal means of celebrating a major holiday or cultural event and in a large public gathering that maximizes public attention. Some communities have long traditions of staging parades, as in the case of New Orleans that has a major parade once every week during the month of May (Sakakeeny, 2010). Other communities, however, have not discovered the power of a parade. They are relegated to seeing parades on television or in some other community, and miss their potential contributions to their own community.

Community practitioners are relatively new to the practice and study of parades, yet they can bring a unique perspective that ties values and social change to practice, bringing an important dimension to this subject's knowledge base that can assist countless other communities in deciding whether or not to stage a parade. There are no obvious aspects related to parades that we envision practitioners gravitating to or away from. Yet there will be aspects related to the seeking of sponsorships and evaluation that may prove challenging. Facilitating parades as a form of community practice provides a natural venue for community involvement.

A successful parade requires a considerable amount of planning and negotiation with various political entities and stakeholders in order to obtain the requisite permits for the undertaking. This task may seem obvious and straightforward. However, in some communities, particularly those with histories of marginalization or that are home to a large numbers of newcomers, dealing with governmental authorities will not generate many volunteers; nonetheless, this is a task that must be completed for there to be a parade.

Parades have drawn the attention of historians, folklorists, urban planners, and social scientists because they represent a very public window into a community or society (Wherry, 2011). Malek (2011, 388), who studied the role and significance of the parade staged by Persians in New York City, one of the 180 annual ethnic parades in Manhattan, finds that this parade has internal and external goals and outcomes, and can be conceptualized as a public diaspora event that can help negotiate Iranian diaspora identities across generation. Unfortunately, helping professionals such as social workers, community psychologists, and counselors have not played a prominent role, although they can play a significant role in fostering the staging of parades.

Where parades are held is an important factor in staging them. Parades invariably take place outdoors because staging them necessitates open spaces to accommodate large crowds and moving floats, bands, and vehicles. Thus, when parades occur is dictated by weather and seasonal considerations, making this event less flexible than fairs and festivals, and this is a key factor in determining which types of events to sponsor and when to do so.

Parades involve movement over an extended geographical area, and they are highly dependent on machinery, such as vehicles and floats and other moving mechanistic devices. Further, parades have a very limited time period in which they can take place, and thus parades are at the mercy of weather conditions. Fairs and festivals can go on for weeks but parades are limited to particular hours and daylight. Yet parades bring dimensions to celebrations that can involve wide media coverage, as well as more participants and spectators, creating or reinforcing community identity and digitizing community histories (Bartoş, Balş, & Berger, 2014; K. Han et al., 2014).

Using websites to promote parades can expand the potential audiences (Castro & Gonzalez, 2014) and be used to predict transportation needs (Pereira, Rodrigues, & Ben-Akiva, 2015), as well as advertise the identities of the communities sponsoring such events (Gillberg & Adolfsson, 2014). In general, you do not have to pay a fee to watch a parade, although there may be fees (entry/sponsorship) to actually participating in the parade. Fairs and festivals, too, can be free but there is an expectation that attendees will spend money buying food (Rong-Da Liang et al., 2013), taking rides, and playing games at a fair; those expectations don't necessarily exist for parades.

Quan-Haase and Martin (2013) address the potential of celebratory events being digitally captured and reaching a broader audience beyond attendees through the use of communication devices and note how this can help to shape perceptions and impressions across a geographical area. The potential to reach beyond the community's borders can be maximized to the benefit of the hosting community sponsoring the celebratory events, as in the case of digital storytelling (Matthews, 2014) and other media forms (Zillinger, 2014).

Parades may appeal to different communities, as in the case of St Patrick's Day parades, bringing forth different identity enactments (Pehrson, Stevenson, Muldoon, & Reicher, 2014). Nevertheless, these very same parades can bring out dissenters and expose tensions within the communities sponsoring them (Scully, 2012).

Conclusion

Parades will never be confused with other events. Activities associated with parades are so different from fairs and festivals, and their uniqueness brings both strengths and limitations. Parades play a prominent role in how nations celebrate. Parades are sometimes a major event, with corresponding national and even international coverage, but they can also be staged on a smaller stage, bringing many of the positive elements associated with such major staging but at the local level. Thus, parades take on highly strategic positions within the constellation of community celebratory events, yet bring all of the logistical and political challenges associated with any major event.

Parades can be successfully tied to other events that can occur at the end of a parade, adding a dimension to this event and extending the celebratory experience into new realms, with complementary goals. This extension of celebration does not come without a price for those working and hosting these events. Nevertheless, parades can have an anchor role in a series of celebrations over the course of an extended period.

SECTION 3

Case Examples of Community Celebratory Events

This section provides case studies of community celebratory events that reflect an international perspective, and facilitate theory integration, to show how these events play an important function within communities and society. These cases demonstrate which forms of events can best meet local goals when grounded culturally and contextually, as well as spotlight how they are evolving in response to local circumstances.

8 Case Example of a Fair: NYC Chinatown's Health Fair

Urban fairs have long traditions of being sponsored by community groups and providing a venue for reaching out to residents in a manner that facilitates interpersonal contact, fosters fun and play, and in the process, accomplishes cultural, social, and political goals. Unlike festivals and parades, fairs can integrate, for example, health and social service efforts directed at "high-risk" groups that pose challenges in being reached through conventional ways, as in the case of newcomer communities.

Any quick Google search of "community fairs" will reveal hundreds of examples of fairs together with health, education, and human services, illustrating the appeal of this event. This chapter highlights an effort at integrating these types of services in the case of a fair; this case was specifically selected for showing the potential this celebratory event holds for transforming the lives of those in the community who are marginalized. The case of New York City Chinatown's Health Fair was selected because it is out of the ordinary and worthy of special attention, and because it complements the San Francisco Chinatown's Lunar New Year Festival, discussed in the next chapter – Chinatowns on opposite coasts of the United States. The Health Fair in NYC's Chinatown shows the evolution of a street fair from inception to the creation of a major urban community institution, the Charles B. Wang Community Health Center, which serves both native-born and newcomer Asians in the community (Hoobler & Hoobler, 2012).

The Charles B. Wang Community Health Center has become an integral part of the social fabric of the community and incorporates key forms of capital associated with being a major asset for a community. Nevertheless, it still sponsors health fairs because their importance to the community has not waned even though the community now has a major institution serving its health care needs. Using health as an organizing theme for the fair allows it to address a multitude of subjects of significance to this community, and in a culturally affirming role. The fair now reaches out to newcomer Asian groups from countries other than China.

The historical context of the Chinese street health fair introduces key sociocultural and sociopolitical elements associated with events, serving as an illustration of a community coming together in a culturally affirming manner (Kline & Huff, 2008). In addition, this case serves as a national role model for other urban communities seeking to reach Asian newcomers through a focus on health, and a desire to expand into other spheres related to political power.

History of NYC's Chinese/Asian Community

Understanding and appreciating current conditions or the situations of newcomers is not possible without a historical context (L.S. Smith, 2001; Rothstein, 2014; Wang et al., 2007; Zhou, 2010). Anderson (2014, 87) addresses "Chinatown" and the social forces that create and maintain this designation: "It is possible, however, to adopt a different point of departure to the study of Chinatown, one that does not imply reliance upon a discrete 'Chinatown' as an implicit explanatory principle. 'Chinatown' is not 'Chinatown' only because the 'Chinese' whether by choice or constraint live there. Rather, one might argue that Chinatown is a social construction with a cultural history and a tradition of imagery and institutional practice that has given it a cognitive and material reality in and for the West."

Carter (2013) addresses geographical forces that shaped how Chinese newcomers gravitated to major cities in the United States and specific neighborhoods within these cities. Daniels (1988, 69–70) points out the close relationship between Chinese migration to big cities and Chinese people in urban centers with established Chinatowns, helping to appreciate how Chinatowns evolved: "Initially, large city meant San Francisco, which Chinese called *dai fou* or 'big city.' By 1940 only 17,782 of 55,030 large-city Chinese Americans (32.3%) lived in San Francisco, with an additional 3,201 (5.8%) across the bay in Oakland. Seven other cities had more than 1,000 Chinese in 1940; there were 12,302 in New York, nearly 5,000 in Los Angeles, and just over 2,000 in Chicago. Seattle, Portland (Oregon), Sacramento, and Boston each had between 1,000 and 2,000 Chinese." Not unexpectedly, today, Greater New York (with a

Chinese population of 735,000), San Jose-San Francisco (629,000), and Greater Los Angeles (735,000) are the three United States areas with the largest Chinese concentrations (U.S. Census Bureau, 2012).

Community celebrations often emerge out of a need for the people in the community to come together and create narratives that affirm their existence and represent vehicles through which they can achieve a wide variety of goals encompassing social, cultural, political, and economic agendas. Newcomer groups often experience social forces that relegate them to unattractive geographical sections (places) within their new homelands. These forces, combined with internal forces that produce a new homeland or community, create distinct ethnic places and spaces (Herzog, 2004). Celebrations become vehicles for marshalling and transmitting various forms of capital to fight social forces that keep these communities segregated, disempowered, and marginalized.

Context is greatly influenced by who is telling or recording the story. The history of Chinese and Asian newcomers to New York City has tremendous parallels with that of other major urban areas of the United States, including San Francisco (see the next chapter), and Toronto and Vancouver in Canada (Mallee & Pieke, 2014; Takaki, 2012).

Foner's (2001) classic book on New York City immigration highlights how different newcomers experience the city even though they share the same destination, transforming the city and being transformed in the process. Each city's unique set of circumstances necessitate a localized grounding of events to take into account how contextual forces give celebrations meaning. The expression "city as context" captures how nuances and social events must be accounted for when telling the story of a newcomer group, and the Chinese of the Lower East Side of New York City are no exception. City context meets historical city of origin context to provide a people's narrative, their hopes, aspirations, trials, and tribulations.

"Chinatowns" can be found in many major cities in Canada and the United States, but particularly on the east and west coasts. These coastal communities, because of their geographical location, have helped shape public perceptions of these communities. New York's Chinatown as an ethnic/geographical community can be traced back over 150 years to 1858, and is considered the oldest Chinatown in the United States, although San Francisco would dispute this claim (Xu, 2013). Chinese newcomers concentrated in residential areas of New York City as well as in certain areas of employment, including laundries, restaurants, and the garment industries, which were integrally connected (Hatton & Leigh, 2011).

Lower Manhattan historically has been a major draw for newcomers to the city dating back to the nineteenth century and a period massive immigration to the United States and New York City, and the Chinese were no exception (Mendelsohn, 2013). By the end of the 1980s it is estimated that over 50,000 newcomers had moved into Chinatown and over 150,000 settled in other parts of the city, most notably Flushing (Queens) and Sunset Park (Brooklyn). Continued migration into Lower Manhattan's Chinatown has spilled over into Little Italy, causing racial tensions between these two different neighborhoods with histories as destinations for newcomers. This population shift was captured by the 2009 American Community Survey, which noted that in Little Italy there were 4,400 foreign-born residents of whom 3,916 (89%) were born in Asia (Roberts, 2011).

Chinatown's expansion has continued this tradition with an increasing number of newcomers moving into Little Italy, making it even smaller in geographical size as its Italian residents either die or move out into other communities. These shifting demographics make community events cross conventional geographical boundaries, and make them that much more important in creating a sense of community based on ethnic, racial, class, and cultural factors.

History of the Charles B. Wang Community Health Center

The health needs of Chinatown's residents meant the neighborhood needed its own facility that embraced both culturally competent and humility goals, and that was willing to fight for community power and rights (Wasik, 2012). In 1970s it was discovered that many of the Chinatown community health needs were going unmet. In an effort to redress this, volunteer medical and nonmedical personnel organized a health fair. They closed off three streets and set up booths for conducting medical tests. Volunteers were required to act as translators at the local hospitals. The Charles B. Wang organization wanted to raise awareness of the special needs that Chinese patients presented as well as meet their medical needs. Social demonstrations followed to put pressure on Gouverneur Hospital to hire Chinese staff. This, in turn, not only increased the quality of health care but also served to create economic capital for the community. It is important to emphasize that health fairs often have an audience beyond those who attend, and in the case of the Chinatown Health Fair, it involved putting another hospital on notice through social activism targeting their institution.

The Health Center started on Mott Street and is the largest medical facility in Chinatown. Yet this institution's significance goes far beyond the provision of health services to an underserved group. This center could not have been possible without significant civic engagement and the tapping of Chinatown's assets. The use of a health fair as a central vehicle for drawing on cultural traditions of bringing people together in a nonthreatening and affirming manner fit well with Chinese traditions (E. Jones, 2014), creating collective memory.

New York City's Chinatown Health Fair showed how community needs were not being met by existing institutions and why they needed their own health center, leading to the development of the Charles B. Wang Community Health Center. Health fair success is more dependent on community engagement and empowerment than on the provision of health services (Knight, 2014). The physical and cultural isolation of this community served to set the context for solidification of power within, although not without sacrifice (Wasik, 2012):

> What this fair showed about the Chinatown community was that it tended to keep to itself and rarely went outside the boundaries of the community when seeking aid. It was always very difficult for the Chinese immigrants to get aid from the city especially because of the language barrier and ... many immigrants where [sic] unable to afford medical assistance. However, for Chinatown to succeed, it needed money and help from the community.
>
> Rather than reaching out to other communities, the Chinatown community banded together to create a powerful platform for the new Medical Center. By 1972, the clinic applied as a nonprofit organization in order to apply for [a] grant from foundations and government agencies. The clinic was growing rapidly and by 1975, thanks to the huge assistance of Jane T. Eng, the Charles B. Wang Center was able to become a Federally Qualified Health Center ... which enabled it to tap into state and federal funding. This allowed for the center to evolve rapidly. By 1979, the Charles B. Wang Health Center had a permanent site on Baxter Street and finally move[d] from its temporary location inside of a church.

Community insularity brings self-reflection and the taking of socialcultural stock, which can be mobilized to create economic and political capital in social changes.

The spark provided by one individual can ripple across a wide expanse to create lasting change in a community (Wasik, 2012):

The Charles B. Wang Center was sparked by the idea of Dr. Thomas (Tom) Tam with the attitude of "Let's be ambitious." Chinatown in the 1970's was going through an abundance of changes especially after the Vietnam War, the Immigration Act of 1965 that ended the discriminatory restriction against Chinese and the civil rights demonstrations and War on Poverty of the late 1960's and early 1970's. Starting in the early 1970's young Asian Americans that felt great pride for their ethnic community of Chinatown organized the Health Fair organized on Mott Street. During this time, the need for medical treatment in Chinatown was immense. The main organizer Dr. Tam worked ambitiously with New York City and Chinatown to obtain access to Mott Street to provide some basic medical needs to the community. After considerable time, the city agreed and the Health Fair was allowed to go through.

The coming together of indigenous leadership and social forces proved to be the right chemistry for the creation of the Health Center.

Sustainability is a hallmark of a celebration evolving with communities changing. In 2011, the Health Center celebrated its 40th anniversary, a testament to sustainability, which is one of the goals of such events. From the Center's website we know: "On July 31, 1971, the very first Chinatown Health Fair was held by a group of volunteers to provide health education and screenings to the medically underserved Chinese community in New York City. Following the tremendous community response, the Chinatown Health Clinic opened its doors that same year, run entirely by volunteer doctors, nurses, social workers, community health workers, and students." This community beacon could not have been possible without the spark provided by the Health Fair, reinforcing the lacunae in health services and the importance of having their own center.

The Health Fair continued to be held, even as the clinic grew in size and scope (Wasik, 2012):

> By the 1990's the Clinic was so popular, that it moved yet again to a larger location on 125 Walker Street where they could start offering specialized care. The health center began having a much larger personal staff and obtaining many more grants and donations from the community. As the community helped, the center also gave back by providing more programs such as a Mental Health Program (known as the Bridge Program), and even expanding to newer locations not only in Chinatown to its current location on Canal Street, but also to Flushing. The entire time, the center still had its annual Health Street Fair.

Few health fairs can succeed to such a high degree, and they only do so when having a fair is not sufficient to trigger social change.

Community Health Fair, New York City Style

Dillon and Sternas provide a succinct definition of a *community health fair* (1997, 1): "A health fair is a community health strategy used to meet community members' needs for health promotion, education, and prevention." Although a health fair was instrumental in the development of the Charles B. Wang Health Center, the popularity of this event has led to its continuation through sponsorship of three health fairs taking into account the changing nature of newcomers and their dispersal across New York City. Two continue to be held in Chinatown and the other is held in Flushing Queens, a neighborhood with a high number of Asians newcomers. A health center and affordable housing started development in 2014.

The importance of the Chinatown Health Fair was such that it continues to be held even though the community has its own health center (Wasik, 2012). The Health Fair gave voice to a community of newcomers and continues to this day. It is noteworthy that the Chinatown Health Fair was so popular that even though it was originally scheduled for one week it was extended to 10 days.

Although the typical health fair has the potential to screen and connect individual attendees with appropriate resources, it often fails to do so (Moore, 2014). That was not the case with the Chinatown Health Fair particularly after it was sponsored by the Health Center. These fairs appealed to newcomers, and offered an important dimension of community control by introducing them to the Health Center for follow-up health visits and providing an opportunity to meet other people where they live.

Mission, Vision, and Goals of the Community Health Center

The Health Center bears the name of a significant financial benefactor, and its mission is as follows: "Our mission is to eliminate disparities in health, improve health status, and expand access to the medically underserved with a focus on Asian Americans" (from the Center's website). In order to achieve this mission, it is necessary to think about how health disparities occurred and the role of social justice in addressing the role of oppressive forces in causing this situation. Their vision "is to strive to be a Center for Excellence by being a leader in providing quality, culturally relevant, and affordable health care and education, and advocacy on behalf of the health and social needs of the medically underserved with a focus on Asian Americans" (from the Center's website).

The goals guiding the NYC Chinatown Health Fair are not highly unusual in their thrust, in accordance with the mission statement above. These goals are the following: (1) provide attendees with an opportunity to be screened for major and common diseases and health conditions in a manner that is culturally syntonic; (2) undertake community education to raise political awareness about employment practice issues at the Gouverneur Hospital (east of Chinatown); and (3) undertake health promotion pertaining to health care and disease prevention. The focus on the hospital and need for these institutions to employ staff reflective of the community's ethnic and racial composition is a perennial theme when community composition changes and the institutions entrusted to serve it lack cultural capital to ensure quality services (Hum, 2004).

These goals of the NYC Chinatown Health Fair introduce social and political considerations and concerns that are typically an integral part of celebratory events. An implicit set of goals involve enhancing the identity of organizations in service to communities and delivering services in a manner that is festive and nonthreatening, and during periods that are conducive to attendance.

Increasing human capital, of which health is a vital segment, and increasing local employment opportunities, brings an asset perspective to this form of event, including an avenue for the undertaking of social activism to achieve these goals (Lavery et al., 2005; Murray et al., 2014). The fair's focus is not simply on "health," but also brings the potential for social activism.

Charles B. Wang Health Services

Health fairs must create an atmosphere that is nonthreatening to attendees and that is free from concerns usually associated with visiting formal institutions, such as concerns about documented status and having proper health insurance (Wasik, 2012), and the success of these types of institutions is measured by their evolution and sustainability: "As the clinic grew though, it concentrated more and more on the immigrant population since for them, this was the only place where they can get medical attention and communicate easily. Also, cutbacks in social programs during the 1980s affected the major hospitals surrounding the

Chinatown area. The Charles B. Wang Health Center ended up being the only place where many people could be treated." The expansion of the Health Center's reach reinforced the central role in the social and political life of Chinatown and its connectedness to the community as a source of power and legitimacy.

The Health Fair played an instrumental role in giving voice to a community of newcomers and continues to do so to this day (Wasik, 2012):

> By the mid-2000's the Clinic has expanded so much, it has now become part of the nation's primary health care safety net ... The Charles B. Wang Health Center now sits comfortably at a large location on Canal Street with over 500 permanent employees and over 42,000 patients. Over ninety percent of its patients were best served in a language other than English. The clinic provides medical treatment to all people both legal and illegal immigrants and citizens. To families with difficult backgrounds, they attempt to organize easier payment plans and ways to apply for government assistance while illegal immigrants may [receive assistance] to obtain citizenship ... The center still remains true to its initial goal that began in the 1970's by providing the best care in a form that best fits the community and empowering the community. The community itself remains relatively isolated and focused on sustaining itself rather than reaching out to other communities in order to mix and grow.

Health fairs can reach significant sections of a community and in the process serve bonding and bridging social capital goals.

Charles B. Wang Center's Unconventional Services

The Health Center provides services that go beyond what is typically associated with health care in the United States, and this flexibility stands as a testament to the vision of its leadership. One such additional service is a book fair and other literacy supports (Wasik, 2012):

> The Charles B. Wang Health Center also has a lot of programs to help increase literacy in the community. Two programs to assist young children are the Reach and Read Program and the other program is Reading is Fundamental. These programs (sponsored by the federal government) supply the Charles B. Wang Health Center with a multitude of books including ethnic books. In order to raise literature awareness not only for the children but the older community as well, Book Fairs are organized in order to assist and maintain interest in reading.

Book fairs and literacy are rarely associated with health fairs, but health fairs lend themselves to creative approaches based on local needs and circumstances, including reaching out to children.

Recruiting and training a cadre of medical staff, too, became a part of the services the Health Center provided to the community (Wasik, 2012):

> Project AHEAD (Asian Health Education & Development) is one of the oldest community oriented programs ... This program, which has been with the Charles B. Wang center since 1975, is designed to provide training and experience for students that are interested in the medical field. This program was initially developed to assist the Asian community since for many Asian Americans; it was difficult to excel for Asians in the 1970's due to the racism in the United States. Now this program is more widespread but still is more focused on the Chinatown community.

Human capital is associated with health, a critical dimension for the Chinese community. If residents have compromised health, they are limited in their involvement in paid and unpaid work (civic engagement) in their lives and communities. Providing training and employment opportunities enhances human and economic capital.

Media: Radio Shows and NYC Chinatown's Health Fair

Media and celebration are integral to event success because they represent a mechanism to engage those who cannot attend an event. The potential of media can go far beyond events. In this case, the introduction of radio programming brings an added dimension to the success of health fairs. Use of community media outlets such as cable television and radio facilitate health outreach and education (Hoberman in Howley, 2005, 133; Wakefield, Loken, & Hornik, 2010) by putting critical media outlets in the hands of communities rather than communication conglomerates. Local sources allow institutions to tailor their message without having it filtered by outsiders, adding to the importance of the Health Center.

The Charles B. Wang Health Center sponsors two different radio shows on two different stations. One show transmits three times per week; the other once per week. These shows address a variety of health-related topics of particular relevance to the Chinese/Asian community and reach important segments of the community that are more accustomed to listening to the radio or who cannot leave their homes to attend workshops or other educational events because of mobility limitations. Radio programming, along with newspapers and magazines, is often used to convey health information and other critical information back to China (Peng & Tang, 2010) and to major cities in the immigrants' new homelands (Kong, 2013).

Conclusion

The image of a fair conjures up memories of coming together in a low-stress and high-fun environment where entire families, as well as neighbors and friends, come together to eat and have a fun reminiscing about old times (collective memory). A throwback to simpler days is clearly one possibility when communities host a fair, particularly one that commemorates an important day in their history, cultural heritage, or major community achievement. Fairs can certainly be devoted to distributing information and providing services related to health and social care needs.

The New York City Chinese Health Fair that eventually led to the creation of the Charles B. Wang Health Center is a model for other communities. The importance of the health fair is such that the Health Center still sponsors health fairs as a means of reaching and mobilizing a community. The center's consumers have diversified from an almost exclusive focus on Chinese patients to include Korean, Malaysian, and Vietnamese people, reflecting demographic shifts in the community. The latter two groups understand and can communicate in Chinese.

Urban health fairs can play a strategic role in communities and mobilize internal resources in search of social, political, and economic goals that can be used in other spheres. These themed fairs can involve local human and educational organizations that normally would not be associated with a celebratory event, increasing the likelihood of connectedness of the residents who would normally not seek the services of these types of organizations. Under the heading of "health" other resources can be integrated, including opportunities to undertake social activism.

9 Case Example of a Festival: San Francisco Chinatown's Lunar New Year Festival

Festivals bring a grandeur that is rarely associated with fairs, although not as great as for parades, creating optics that lend themselves to extensive media coverage (Murphy, 2014; Oliveirinha, Pereira, & Alves, 2010), and the possibility of a large crowd and a high-energy ambiance. Large crowds bring a high degree of impressionability that limits the depth of interaction and connection between participants, yet can serve broader social and political goals.

A Yelp review by a former participant (Kitty K) of San Francisco Chinatown's Lunar New Year Festival brings to life many of the key elements, rewards, and tribulations associated with festivals:

> Prior the parade, the streets of Chinatown are closed off with the typical vendors you see during all the Chinese festivals. Going here is also an excuse for me to go to all my favorite shops in Chinatown. The parade brings more people, however, so of all the festivals held in Chinatown, this is the most packed … Parking is also EXTREMELY difficult to find and traffic is pretty bad, so be sure to have patience. Still, love all the color and the energy from everyone's excitement. Definitely something to go to at least once. Note: Be weary [sic] of your purses or pockets though. Since it's so crowded you can get pickpocketed easily. I've had a few friends have money stolen from them that way.

Festivals are best understood when viewed within a localized sociocultural and sociohistorical context and a flexible view of how they should be manifested in order to understand their evolution and why they have achieved a high level of sustainability in urban communities. Although few people would argue against fun and games, when they go wrong, the consequences can be significant and long-lasting.

Bowmile (2009, 3) identifies key findings pertaining to shaping and sustaining Toronto festivals, although they are also applicable to other major urban centers:

> Festivals have come to serve cultural and economic functions in the city.
> The Word on the Street has actively searched for ways to situate its exhibits within the context of Toronto's historic, ethnic, and contemporary cultures.
> Festivals reappropriated public spaces, which allow festivalgoers and festival exhibits to interact uniquely with urban settings.
> Festivals such as Word on the Street have democratized culture by making it more available and accessible for public consumption.
> Festival growth has contributed towards improving global perceptions of Toronto as a cultural capital.
> The City itself needs to do more from a financial perspective to champion the city's culture. This is because it remains too dependent on contributions from other, typically private sources.

These findings validate the central thrust of this book on why celebrations are important in capturing and affirming urban places and spaces.

This chapter focuses on one racial group (Chinese), event (Lunar New Year), and city (San Francisco), and follows an event's evolution as it increased in size, significance, and shifted its goals as the experience and context of the Chinese community changed over the past almost seven decades. Anderson (2014) discusses the emergence of Chinatown and the forces that shaped it. These communities celebrated cultural traditions, evolving to reinforce a Chinese community identity within and outside of the community.

This case study of San Francisco's Chinese community taps a racial group with a long-standing tradition and reputation of staging festivals, and this tradition has followed the Chinese as they have settled across the world, and particularly in major cities of Canada and the United States. A specific focus on one racial/ethnic group allows an in-depth understanding of the sociocultural role, and the importance of celebrations in helping a community's newcomers meet essential sociopolitical, socioeconomic, and sociocultural goals in order to survive and thrive in their new world environment.

This chapter consists of five sections, discussing: (1) the modern-day Chinese community and San Francisco, (2) the history of Chinatown,

(3) the history of the lunar festival history, (4) the Miss Chinatown beauty queens and a showcase of traditions, and, finally, (5) major organizational and community themes. Each of these sections focuses on a specific aspect of the celebration, although there is an overlap, illustrating the multifaceted (analytical and interactional) dimensions of festivals within one ethnic community in the United States, and how ethnicity and culture can be assets rather than liabilities.

The Modern-Day Chinese Community and San Francisco

San Francisco's Chinese community is best appreciated against a historical backdrop of immigration from southern China to the United States, which started over 160 years ago (Wong, 1998). The social, political, and economic circumstances that awaited the Chinese immigrants helped shaped their responses at achieving stability and a future in their new homeland (Daniels, 2004). San Francisco's proximity to China made it an ideal port of entry into the United States (Ma, 2014). The long distance and conditions that led the Chinese immigrants to leave China made it necessary for them to think of San Francisco as their new home and their destiny.

Wong's excellent book *Ethnicity and Entrepreneurship: The New Chinese Immigrants in the San Francisco Bay Area* stresses adaptive qualities and adaptive views as assets. Small businesses and their owners have been immensely influential in the establishment of celebratory events. The composition is dynamic and the evolution of this community is complex and goes far beyond geography to encompass social, economic, and political dimensions. Nevertheless, it is not so complex that one cannot offer an explanation and appreciation for why this event has evolved in the manner that it has.

San Francisco's Chinatown Lunar New Year Festival is best understood against a present-day backdrop. San Francisco's Chinatown is home to the largest Chinese community outside of Asia, and claims to be the oldest Chinatown in North America (Yung, 2006). (New York City, too, makes a claim on being the oldest, as noted earlier.) This neighborhood covers a one-mile stretch located downtown, between the financial district, affluent Nob Hill, and the historically Italian district of North Beach, spanning approximately five zip codes; the one-quarter-square-mile area bordered by Kearny Street and Jones Street is regarded as the heart of Chinatown.

This area is home to approximately 14,000 residents, which is a relatively small number of residents in a major metropolis, and it has been a tourist attraction for well over a century. Chinatown is a significant tourist destination estimated to attract more visitors annually than the Golden Gate Bridge (SF Gate, 2014), yet despite its reputation as a tourism powerhouse, the socioeconomic profile of its residents has remained relatively consistent throughout its 160-year history. Residents consist of Chinese and other Asian Americans, most of whom are low-income, employed in the service industry, and/or economically and culturally isolated from the broader community (Pamuk, 2004).

The neighborhood remains a working-class community despite the growing affluence of multiple generations of Chinese Americans in the Bay Area. Chinatown remains a highly transient neighborhood and entry point for Asian immigrants to the United States (Pamuk, 2004). Chinatown homeowners, on average, live in overcrowded conditions and longer-term residents are generally low-income, elderly, and/or linguistically unintegrated into mainstream society. Today, many Chinese American communities are found throughout San Francisco, and residents are clustered by economic status; over the past 50 years, there have been notable migration patterns of more affluent Asian American families toward western portions of the city, and away from Chinatown (Pamuk, 2004).

Many smaller, unofficial "Chinatowns" have sprouted throughout San Francisco allowing higher-income Asian American residents to invest in the economic social development of neighborhoods, such as the Sunset and Richmond districts, rather than downtown Chinatown. Chinatown's historical significance remains to this day and has been a subject of numerous scholarly articles and books, not to mention books on tourism signifying how cultural has been commodified as economic capital.

History of San Francisco's Chinatown

It is essential to ground the Chinese New Year celebrations within a historical context in order to appreciate how social forces shaped the evolution of this community (Yeh, 2008). Early Chinatown residents were Chinese immigrants from southern China's Guangdong Province who emigrated in record numbers between 1850 and the early 1900s (Jorae, 2009). Most initial immigrants were male and of Hoisanese ethnicity, and worked in the transcontinental railroad, as mine workers and as laborers during California's Gold Rush period in the mid-1800s (Jorae, 2009). These jobs were plentiful and newcomers were restricted from pursuing other types of employment because of racial discrimination.

158 Case Examples

The first Chinese inhabitants settled on one block of Sacramento Street in the 1850s, and by 1885 the community had expanded to over 12 city blocks (Yeh, 2004). The completion of the transcontinental railway in 1869 was accomplished with the labor of over 12,000 Chinese migrant workers in California (Choy, 2012). As the Chinese community increased in size, residents found work in the service industry, opening laundries, restaurants, and finding employment as domestic workers and laborers in the textile and garment industries (Choy, 2012; Yeh, 2004). As for its appeal as a tourist destination, Chinatown merchants have long capitalized on the "exotic" image of East Asian culture to attract visitors nationwide (Choy, 2012). This is an example of cultural capital transforming into economic capital, although certainly not without its limitation of feeding into stereotypes.

The Chinatown seen today is a reconstruction of the original settlement of the late 1800s (Rast, 2007). Chinatown was rebuilt by local Chinese business owners after its complete destruction in the historic 1906 earthquake and in response to city plans to move Chinatown to an undesirable location (Yung, 2006). A result of the reconstruction is a tourist-friendly style including a cleaner appearance and exaggerated "Oriental" architecture (ibid.). After the reconstruction many of the local businesses advertised goods and services through colorful and exotic posters, placards, and signage (Yeh, 2004), giving the community the unique identity still seen today, and explaining how such a relatively small community can build on its past and wield such an influence external perceptions.

History of the Lunar New Year's Festival

Unlike many other celebrations that are held on the same month and day of a year, the Chinese New Year is a lunar festival and so when it is staged varies because that date of the moon's phases varies from year to year. Although the celebration dates are determined by the day of the week, the parade is always staged on a Saturday to maximize attendance and publicity. The parade is usually held two weeks after the initial day of the Chinese New Year, with over 100 groups participating.

The 2014 celebration consisted of the following four formally sanctioned celebratory events that occurred over a weekend period: (1) the Chinese New Year Flower Fair, where families could purchase traditional plants and flowers to decorate their homes and/or give as gifts; (2) the Chinatown Community Street Fair, which featured traditional arts and performances; (3) the Miss Chinatown USA Pageant; and (4) the Chinese New Year Run, a 5-and 10-kilometre race with funds benefiting the Chinatown YMCA. Museums and other institutions sponsored special activities celebrating this holiday. There was a self-proclaimed largest treasure hunt in the United States, under the banner of the Chinese New Year – all this illustrating how one major event can be coupled with various other events.

San Francisco's Lunar Festival is considered to be the largest in the United States and arguably the most famous (Sievert, 2006). History shows that San Francisco's early Chinese New Year festivals were a direct response to racial tension caused by Cold War political and economic conditions. The first festival leaders were influential in San Francisco's Chinatown politics and considered themselves "cultural brokers" between Chinese Americans and the community at large (Yeh, 2004). The festival began in the 1950s as a concerted effort to demonstrate patriotism and anti-communist convictions as well as to promote "Chinese American gendered ethnicity" (395) and distance the community from "Red China" (Yeh, 2002, 2004).

The political context of this event took on more obvious connotations in 1979, when the United States formally recognized the People's Republic of China. The Lunar Day Celebration witnessed the appearance of that country's flag and a disappearance of Taiwan's national flag, representing a shift away from Taiwan in response to a changing US foreign policy (Yeh, 2009).

Early Chinese cultural festivals in California (starting in the 1880s) largely ignored religious aspects of the holidays being celebrated, but instead emphasized the aspect of cultural exoticism, and this continues in modern-day Chinese New Year festivals in San Francisco.

Rath (2007, 2) addresses how the evolution of the Lunar New Year celebrations has changed over the years in response to sociocultural changes: "Originally, the Lunar New Year celebrations were a matter of locals only, an occasion to meet, greet and experience and strengthen feelings of belonging to the Chinese community, but they increasingly involve a broader public." Sustainability and evolution are closely associated if celebratory events are to maintain relevance.

The first organized San Francisco Lunar New Year festival started on 15 February 1953, with prior celebrations being private and family-oriented (Yeh, 2004). The festival was sponsored by the Chinese Chamber of Commerce (CCC) whose stated primary purpose of the festival was to increase tourism in Chinatown and promote patronization of local Chinese-owned businesses (Yeh, 2002). A related goal was to alter and guide public perception and stereotypes of the San Francisco Chinese American community, as for nearly a century prior to this, the community faced overt racism through the passage of several pieces of legislation aimed at limiting Chinese immigration and economic opportunities (Choy, 2012; Yung, 2006). In the 1950s, Chinese Americans still faced severe social, political, and economic persecution (Yeh, 2002; 2004). During this period, Chinese Americans were treated with suspicion as both McCarthyism and the Korean War had traumatic impacts on this community. In San Francisco, the relationship between tourists and the local Chinese American community was "one of suspicion" and the atmosphere was fearful, and intermittently a subject of discussion (Rast, 2007; Yeh, 2004).

And so the festival was created as a symbol to influence the perceptions of Chinese Americans and shift these perceptions away from stereotypes of a communistic people to that of a "docile, exotic, and feminized minority" (Yeh, 2004, 398). The goal was to create a "safe minority" to combat racial tension and attract mainstream "consumers" of the culture by playing up exotic cultural aspects. The parade was an attempt to create a unique Chinese American ethnic identity in an unobtrusive and unthreatening manner. Early festival organization efforts were a means to reduce racial tension by leveraging the American fascination with "Orientalism" (Yeh, 2004). Women and children attending the festival were encouraged to dress in traditional attire, promoting a gentle cultural integration and harmonic balance of "East meets West."

The first festival included music and art shows, street dancing, and a fashion show, followed by the evening parade which is still widely attended and receives extensive media coverage locally and nationally (Yeh, 2004). As popularity soared over the next 10 years, other events were added including a smaller daytime parade, Chinese movies, and a Miss Chinatown beauty queen coronation – all of which increased the sustainability of these events and created a significant cultural theme that is only possible when there are multiple celebratory events complementing each other.

Miss Chinatown Beauty Queens and a Showcase of Traditions

Beauty pageants, although not without controversy because of how they objectify women and reinforce stereotypes, too, are community events, and can be truly community centered rather than mega-events. Pageants are often a part of broader event festivities usually marking a festival or parade, as in the case of San Francisco's Chinatown parade and festivities. Starting in 1958, Miss Chinatown USA aimed to change the public perception of Chinese American women, which at the time was limited to one-dimensional stereotypes fueled by the mainstream media. The New Year Festival paired with the success of Miss Chinatown USA promoted the community as a tourist destination and encouraged the Chinese American identity as lawful, average American citizens (Yeh, 2002).

Major Organizational and Community Themes

San Francisco's Lunar Day Parade incorporates many of the key elements in the staging of a parade and additional ones, too. The planning of such a major celebratory event raises prodigious complexities for a sponsoring organization(s) and community, but the rewards to a community can be incalculable when they succeed and evolve over decades. An ability and willingness to respond to changing demographics and economic and political forces helps ensure sustainability, as in the case of the Lunar Day Festival in San Francisco, and its economic benefits extend beyond the immediate community to the city as a whole.

San Francisco's Lunar Day Festival illustrates how disparate groups sharing ethnicity can overcome opposition from within and outside of the community, tightening the community, and bridging gaps between groups that historically have had conflict and tension with each other (Yeh, 2009). On the surface, the Chinese community may appear homogeneous but it is, in fact, highly diverse, and this diversity can be expected to continue for future generations.

Another major theme of the Chinatown celebration is its ability to generate revenue by way of tourism and increased general commerce, and for the city, too, thus improving Chinatown's image away from that of a low-income community. There is now a website specifically devoted to this community and the events that it sponsors throughout the year, and it illustrates how this celebratory event is marketed (http://www.sanfranciscochinatown.com/events/chinesenewyearparade.html):

> Chinese New Year is a two week Spring festival celebrated for over 5,000 years in China. The San Francisco Chinese New Year celebration originated in the 1860's during the Gold Rush day and is now the largest Asian event in North America as well as the largest general market event in Northern California. The celebration includes two major fairs, the Chinese New Year

Flower Fair and Chinatown Community Street Fair. All the festivities culminate with Chinese New Year Parade.

> Named one of the world's top ten parades, Chinese New Year Parade in San Francisco is the largest celebration of its kind outside of Asia. Over 100 units will participate in the parade, many of the floats and specialty units will feature the theme of this year's Chinese zodiac sign. Nowhere in the world will you see a lunar New Year parade with more gorgeous floats, elaborate costumes, ferocious lions, and exploding firecrackers. Some of the parade highlights include elaborately decorated floats, school marching bands, martial arts group, stilt walkers, lion dancers, Chinese acrobatics, the newly crowned Miss Chinatown USA and the Golden Dragon. The Golden Dragon is over 201 feet long and is always featured at the end of the parade as the grand finale and will be accompanied by over 600,000 firecrackers! The Golden Dragon was made in Foshan, a small town in China. The Foshan dragonmasters formerly made all the costumes for the Cantonese opera, and the Golden Dragon bears many operatic touches, such as the rainbow colored pompoms on its 6 foot-long head. It is festooned from nose to tail with colored lights, decorated with silver rivets on both scaly sides and trimmed in white rabbit fur. The dragon, made on a skeleton of bamboo and rattan, is in 29 segments. It takes a team of 100 men and women to carry the Golden Dragon. This is also considered an honor to be chosen for the grand finale. Rain or shine, come watch the parade!

The description appeals to a wide range of audiences and leaves little doubt about the spectacle.

Celebration was and is a major catalyst in changing San Francisco's tourism landscape, and has resulted in many improvements (not just limited to infrastructure) to the Chinatown community and the city, but it has not addressed major Chinese stereotypes. An opportunity to educate the public about this community is compromised or sacrificed in the interest of economic, political, and social goals.

Conclusion

The San Francisco case highlights how celebratory events perform important functions within the community, including how one event can lead to others and an expanded time period during which they occur. Sustainability is one of the community benefits. The shaping of community identity, too, results as in the case of San Francisco's Chinese community and far beyond the boundaries of that city to a national and international sphere. From a tourism perspective, the branding that has transpired is an event organizer's dream.

The success of San Francisco's Lunar Day celebrations has resulted in a sociocultural price to a group and their community. The political forces leading to the creation of this event continue to this day. The San Francisco Chinatown Lunar Day celebration, however, can be viewed as an event that unites Chinese communities around the world, bringing recognition to this community's presence among the world's major cities.

10 Case Examples of Parades: Toronto's Portuguese, Hispanic/Latino, and Pride Parades

This chapter takes a different perspective compared with the two previous chapters. Instead of covering one festival or fair, the focus is three separate parades (Hispanic/Latino, Portuguese, and LGBTQ) that take place in one city (Toronto), facilitating an appreciation of how different marginalized groups have conceptualized and staged a particular form of celebratory event sharing a similar urban public space. Addressing three parades does not come without its shortcomings. The proverbial debate of breadth versus depth highlights this dilemma.

This contrast in parades creates an appreciation of how a celebratory event can share significant similarities and differences regarding goals and outcomes when sponsored by different groups, and all within the same urban geographical context and time span (summer). There are often many other celebratory events tied to parades, as in the case of Toronto's Pride Week, which covers a 10-day period and involves many different celebratory events including a parade, this serves to complicate an analysis but also highlights the interrelationship between different forms of events.

The study of national days (Portuguese and Latino/Hispanic) represents one of the latest and most promising trends in scholarly research on national identities of newcomer groups (Elgenius, 2011; McCrone & McPherson, 2009), opening up for interpretation and a deepening understanding of how fragmented populations of newcomers perceive national celebratory events in their new homelands, and how these experiences shape goals and how parades unfold (Leal, 2014). National days will only increase in importance in a highly globalized and urbanized world, and parades are often a favorite way of celebrating these days.

The annual Pride Parade, in turn, introduces sexual identity and intersectionality as points of celebration and rallying for various social justice themes that may, or may not, be directly related to LGBTQ issues but represent views that benefit from the wider public exposure afforded by a major parade. Although this discussion addresses each of these events as distinct entities, these three parades are not mutually exclusive since participants overlap. Each parade unfolds with a keen appreciation of sociopolitical contextual forces with deep historical roots within Toronto.

Toronto's Ethnic Neighborhoods and Cultural Heritage

Any word association of ethnic neighborhood will enlist a wide range of responses, particularly those related to food, music, languages, dance, and festivities. Celebrations bring these elements under a singular setting. Toronto has developed an international reputation for its celebrations, as a destination city for newcomers (Ingalls, 2012), and for its LGBTQ community, although not without challenges and tensions (Munson & Chetkow-Yanoov, 2014; Reitz & Lum, 2006). Toronto is not unique on the world stage for ethnic celebrations by different groups using the same type of event to meet their goals (Goonewardena & Kipfer, 2005). Cultural events represent a community's quest for cultural recognition and political empowerment (Premdas, 2004).

Almost half of Toronto's population is foreign-born (Ginieniewicz, 2010). Toronto's rich cultural heritage (stemming from its status as a major port of entry) has resulted in numerous celebrations, not restricted to parades, reflecting this diversity in music, art, and theater, for example (Mercier, 2007–8). The Italian Parade (Stanger-Ross, 2006, 2010), Chinese New Year (Schiller, 2012b), the Trinidad and Tobago Festival (Isaac-Flavien, 2013), and Anthony Day in Little Italy (Stanger-Ross, 2010) are just four examples (Ruprecht, 2010).

Toronto's ethnic neighborhoods have different histories and challenges, but they all see value in conducting celebrations to achieve both internally and externally oriented goals. McClinchey (2008) reports four case studies of Toronto neighborhood ethnic cultural festivals and identifies key considerations – politics and image, social identity and representation, cultural authenticity, and neighborhood differentiation – related to urban place making and festival promotion. These case studies show the value of this method for providing community narrative.

Toronto's ethnic neighborhoods have undergone dramatic demographic changes since they were initially settled by different groups

(Hackworth & Rekers, 2005). The original Chinese neighborhood, for example, has changed as non-Chinese Asian groups have moved in, with out-group migration of the Chinese to other sections of Toronto and its suburbs (Fong, Luk, & Ooka, 2005; Phan & Luk, 2008). The Latino community, too, has undergone changes due to gentrification by white, non-Latinos.

Bramadat (2004, 91) summarizes how celebratory events responded to demographic changes in Canadian cities: "It is clear that these events can continue to address the shifts we are witnessing in Canadian cities. By providing a context in which people can challenge stereotypes, engage in dialogical identity formation, reappropriate popular culture for their own use, and explore foreign (yet nearby) cultural and physical terrains, these events continue to serve a useful role in the Canadian urban conversation." Ethnic celebrations are a window for understanding a neighborhood's ethnic composition and evolution.

Portuguese Day Parade in Toronto

The Portuguese in Canada and Toronto

Immigration after the Second World War is considered to be the primary engine of economic, social, and cultural change in Canada, with its manifestation most profoundly felt in major urban areas and, since the 1960s, in the immigrants with non-European origins (Teixeira, Lo, & Truelove, 2007). Estimates show that approximately half of all Portuguese people in Canada reside in Toronto (Rocha-Trindade, 2009). Entrepreneurs are of tremendous importance because they sponsor community celebrations, and small businesses in Toronto's Portuguese community also are a source of community leadership (Kaplan & Li, 2006; Teixeira, 2006).

The Portuguese community is best understood from a historical backdrop of the arrival and dispersal across Canada of Portuguese immigrants. The 1950s marked the beginning of this immigration due to massive political upheavals in Portugal. Zuev and Virchow (2014) discuss the importance of the Portuguese Day celebration in Portugal and why this event in the immigrants' new homelands takes on such importance for bringing them together as a community. The migration of the Portuguese to Toronto did not signify a unified community resulting from a political divide that could be traced back to Portugal (Fernandes, 2010).

Higgs and Anderson (2013) provide an important summary of the various waves of Portuguese-speaking people coming to Canada, and their settlement patterns across the country: "From a trickle in the 1940s (some 200), Portuguese immigration to Canada increased rapidly after 1953. Immigrants arrived from the Azores (comprising 70% of Portuguese immigration to Canada) and Madeira archipelagoes and from continental Portugal. Many of the 1950s arrivals were recruited to work in rural and isolated locations in Canada, but soon established themselves in the larger cities ... By 2006 the number of Portuguese in Canada was estimated to be ... 410,850." As Toronto's Portuguese community increased, so did intragroup diversity and efforts to celebrate their roots and new beginnings in Canada, creating greater challenges and rewards surrounding events.

Toronto has a long history as a port of entry for newcomers (Nunes, 2011), and Teixeira (2007, 1) provides historical and demographic overview of Portuguese immigration and its evolution over the past several decades: "Portuguese immigration to Toronto began in the early 1950s, and peaked in the late 1960s and early 1970s. As of the 2001 Canadian Census, 357,690 Portuguese lived in Canada (total ethnic origin). The Toronto Census Metropolitan Area is home to the largest concentration of Portuguese (171,545) in the country. The majority of this group (96,815) lives in the City of Toronto, of which 12,075 reside in Little Portugal, the historical core of Portuguese settlement in Toronto."

Harney (2006) addresses urban space politics and illustrates three ways (a range of scales, temporal duration, and purposeful collective expression) Italians in Toronto claim and shape their neighborhoods; these forms of place-making have implications for the Portuguese and other newcomer groups as well. Monuments represent a physical reshaping of urban space and are worthy of further attention as an activity with a natural evolution tied to events.

Little Portugal, Toronto

Little Portugal is an economic, political, social, and cultural center and home to many of Toronto's first Portuguese generation, playing an influential role in the staging of the Portuguese Day Parade, an event that unites multiple generations within and outside of Little Portugal, while serving to educate residents of Toronto about this community's history.

Little Portugal is a neighborhood in transition (gentrification) as second and third generations move out to other parts of Toronto and the suburbs, increasing housing prices and resulting in financial gains for those selling their homes, but to a different group of new residents coming in. This demographic shift has caused tensions between long-time residents and new residents as typified by the following quote reported in Teixeira (2007, 5): "In June at the Portuguese Parade ... a gentrifier complained because we were making too much noise. This is the conflict we have now ... we are very happy when we sell our houses for $750,000 but we get upset when these guys come with their decibel meters, measuring the noise ... two worlds, it seems." These tensions get accentuated when there is a shifting demographic of new residents with higher socioeconomic status and formal education move into the community. Celebrations such as the Portuguese Day Parade hold little symbolic meaning to the new residents, further highlighting their differences from long-term residents.

Portuguese Day Parade

Parades sponsored by newcomer groups have a long history and provide insights for assessing a community's assets, struggles, and dreams. Parades, although significant in their own right, can be part of a constellation of celebratory events rather than a singular event, heightening their impact as well as the impact of the other events. Hatfield (2013), a local newspaper reporter, provides a good description of Portuguese Day and its evolution from a celebration introduced by Portugal's democratic republican government that wanted to secularize the Portuguese community (they chose June 10 as the date of the celebration to coincide with the date of death of Luis de Camoes, a sixteenth-century poet "who shaped much of the Portuguese language and literature") to a holiday supported by the dictator Estado Novo as a "propagandistic, ideological and militaristic celebration" to events for diaspora communities. In Toronto, the first Portugal Day festivities were organized by the head of the Portuguese parish of the Catholic church. Participants later demanded a more democratic organization to the festivities; the Alliance of Portuguese Clubs and Associations was formed in 1986 and has been responsible for organizing Portugal Day festivities in the Toronto area since then.

> The commemorations of Portugal Day have grown over the years to become the largest public gatherings of Portuguese immigrants and their descendants in Canada. In Toronto they now include a large program of activities that span the month of June, and are highlighted by the parade of community associations and businesses along Dundas St West ... The city councilor for the area known as Little Portugal, Ana Bailao, said Portugal Day, today,

is not just a celebration of the history and heritage of the Portuguese community, but also of Toronto ... Bailao has participated in Portuguese Day festivities for many years as well as having been the director of the organization that plans the festivities for one year. She said the celebrations and significance have changed as the community has matured. "For the first generation there was a huge need to connect to your roots," she said. "Now it is more a matter of pride and also showcasing to the second generation. It is more about what it is to be a Portuguese Canadian." (Hatfield, 2013)

The parade's evolution has resulted in the embrace of a broader audience and a wider set of goals for Toronto's Portuguese community, reflecting how sustainability has resulted because of this expansion and incorporation of a new generation.

The Portuguese have a long tradition of public celebrations, so bringing these traditions to a new country is not a surprise and merely represents a continuation of a centuries-old tradition, although not without changes to reflect current issues and considerations (Melo & Robeiro, 2014). The significance and symbolic meanings of these celebrations has undergone a transformation taking into account the evolution of uprootment and how it has altered their lives and connections to the homeland, while introducing a new sociocultural context in their adopted homeland. Celebrations are a bridge between old and new worlds and different generations.

Leal's (2014) analysis of Toronto's Portuguese community through a lens focused on the Portugal Day Parade stresses a multifaceted perspective and the importance of contextual grounding of this celebratory event. West central Toronto's "Little Portugal" area has played a historical role in welcoming this newcomer group, and sponsoring celebratory events (Murdie & Teixeira, 2011; Siemiatycki & Isin, 1997; A. Smith, 2014). Place, space, and celebrations are integrally connected, grounding these events within a local context essential to fully understand the symbolic and practical meaning they have for those sponsoring these celebrations.

Hispanic Day Parade in Toronto

Toronto's Hispanic/Latino Community

Southwestern Ontario historically has been home to many Latin American newcomers (Wilson-Forsberg, 2014). Toronto's Hispanic/Latino community is relatively new, consists of people from over 20 different countries (Veronis, 2010), and is heavily concentrated in the west end, or the "Western Immigrant Corridor," which is often referred to as the "Jane and Finch" intersection (Veronis, 2006; Gibson-Wood & Wakefield, 2013). This Toronto section has a disproportionate number of low-income residents and public housing. This community has struggled to find its place and space in this city's geography, increasing the importance of a celebratory event such as the Hispanic Day Parade.

Hispanic Day Parade

The Hispanic Day Parade provides an opportunity for the articulation of a Latin American identity in Toronto (Gilmartin, 2008), and for the Latino community to educate the broader community about the Latino experience and culture (Ehrkamp, 2008). Their narrative needs to be shared, and the parade is a vehicle for the sharing of their experiences, dreams, and challenges (Racine, Truchon, & Hage, 2008); this parade is a revival of an annual festival (Multicultural Festival) that was held in that community during the 1987 to 2003 period (Veronis, 2006).

The initial parade/celebration, in 2001, was sponsored by the Hispanic Day Parade/Super Latin World Arts Festival, Inc., and it attracted between 30,000 and 50,000 participants. The festival includes entertainment, folklore, music, dance, and traditional Latin American food, as well as a beauty pageant contest known as Miss Hispanidad. Local Latino small businesses play an influential role in the staging of this parade and festival (Veronis, 2006). Over 20 contestants participated and over two dozen floats took part representing different countries.

Unlike typical parades which take place in the downtown core of a city, the Canadian Hispanic Day Parade (CHDP) is staged within the Jane and Fitch neighborhood of Toronto (Veronis, 2006, 1660–1):

> The route of a parade is filled with politics and symbolism ... In the case of CHDP, the choice of its location rather than its route represents a political claim ... Whereas most parades and Latin American events are held in downtown Toronto, CHDP takes place in a suburban and low-income area with significant numbers of visible minorities and disadvantaged groups: over 60 percent of the population is immigrant and 20 percent of the residents are newcomers (arrived in 1996-2001) ... Moreover, Jane and Finch [neighborhood] is experiencing significant disinvestment from the

state ... Finally, the neighbourhood suffers from negative stereotypes and media images that stigmatize it as one of Canada's worse ghettos.

The political socialization of Toronto's Latino community does not occur simultaneously due to diversity within this group creating different political cultures, making finding place and space arduous (Landolt & Goldring, 2009). Belonging (social, identification, and emotional) takes on paramount importance for uprooted groups and public spaces facilitate the articulation of this identity (Blunt, 2007; Yuval-Davis, 2006).

Veronis (2007, 455), a leading scholar on Toronto's Latino community, addresses the importance of Toronto's ethnic groups claiming a place and space they can call their own: "The case of Latin Americans' struggle for belonging in Toronto serves to reflect on how and why new immigrant groups today (re)construct collective identity spatially. I argue that immigrants strategically essentialize their identities in and through place in order to make themselves visible and their voices heard. Ethnic places represent sites of resistance and creation where immigrants construct their own subjectivities while also redefining dominant notions of inclusion and citizenship. Although locally grounded, these new immigrant identities remain fluid and engage with multiple forms of exclusion."

She goes on to quote a Latino Toronto councilor candidate's assessment of Toronto's Latino community: "[The] situation is simply sad; the [Latin American] community ... is one of the most orphan communities ... in [Toronto] ... [We] don't even have a place where to dig our own grave basically. If there is need to get together ... a meeting ... there is no place. We have to be looking for a basement ... for a recreational centre to give us a room ... If there is a social or cultural event, we do not have a place where ... we can present what we have ... [It] is sad and it is a reality."

The following quote by the president of the Canadian Hispanic Day Parade Committee (Toronto) shows the role and importance of having political goals manifested through the staging of a parade (as quoted in Veronis, 2006, 1653): "[The Canadian Hispanic Day Parade] is a way of giving a blow to the government, of saying Latin Americans are doing ... something. The community is at a low level and the government is not listening. But with the parade and the participation of local politicians, the government is forced to pay attention ... They see that the parade is well organized and this goes against the ... negative images that the government has about Latin Americans." The political agenda is multifaceted and challenging to carry it out.

Toronto's Pride Parade

Toronto's LGBTQ community introduces the role of pride parades in Toronto. Brickell (2000) argues that public space has been constructed as "heterosexual space" through making heterosexuality in public unproblematic and normative, and by policing LGBTQ identities by subtle or overt means. Pride parades serve to reclaim urban public spaces and voices, and this process of liberation is facilitated through the engagement of crowds.

The increased use of parades to bring attention to a nation or city's LGBTQ community is illustrated in the case of Toronto's Pride Parade, which has expanded and evolved to include other forms of celebration, and an extended period of coverage, helping to ensure its sustainability. It has evolved from a single day event to an entire week (Pride Week), with parades, festivals, and other events held during this period. Many local businesses count on the revenue generated from this week to help them make profits for a year, which shows the economic importance of this event (Hains, 2014).

The increased likelihood of sustainability must not be lost. It would have taken a vivid imagination back in 1994 to have accurately forecasted that an event that started out seeking to send a social and political message would eventually evolve to have a major economic impact on Toronto (Kallen, 1996). In 2013, Pride generated an estimated $286 million. Burgess (2011) discusses Toronto's LGBTQ community's quest for social justice through an analysis of its Dyke March or Parade and why it is so important to capture this perspective:

> I argue that the Dyke March is a complex, complicated and contradictory site of politics, protest and identity. Investigating "marching dykes" reveals how the subject of the Dyke March is imagined in multiple and conflicting ways. The Toronto Dyke March is an event which brings together thousands of queer women annually who march together in the streets of Toronto on the Saturday afternoon of Pride weekend. My research examines how the March emerged out of a history of activism and organizing and considers how the March has been made meaningful for queer women's communities, identities, histories and spaces ... I argue that the Dyke March is an event which is intentionally meaningful in its claims to particular spaces and subjectivities ... Although my analysis is focused on Toronto as a particular site, it offers insight into broader queer women's activist organizing efforts and queer activism in Canada.

The politicization of sexual identity is integral to public demonstrations for LGBTQ groups, and this goal can be facilitated through public events such as parades (Graff, 2010). Toronto's Pride Parade opened up the door for other social issues, as evidenced by its position against apartheid in Israel (Elia, 2012; Hanson, 2011; Hoxsey, 2012). Hudes (2014, 1) reports on the success of this effort on the 2014 parade, which was not without controversy, illustrating how intersectionality can manifest itself in a pride parade:

> For the third straight year, Queers Against Israeli Apartheid will march in the Toronto Pride parade, but this time with little public opposition from city councillors and pro-Israel groups. In past years, members of council, including Mayor Rob Ford, threatened to cut Pride funding pending the group's participation, or ban QuAIA from marching altogether, feeling the phrase "Israeli Apartheid" was discriminatory. But a 2012 report by Toronto's equity, diversity and human rights department concluded that the phrase didn't violate the city's anti-discrimination policy, and a dispute resolution committee later ruled that QuAIA's attendance at the parade didn't discriminate against the Jewish community either. QuAIA has since marched in 2012 and 2013 amid further criticism and calls to ban it from the parade.

Introduction of other social issues have occurred, too. Toronto's Pride Parade has sought to achieve zero waste and carbon neutrality by 2015 (Dodds & Graci, 2012). Intersectionality can be found in Toronto's Pride festivities, too, as in the case of Latina lesbians (Jiménez, 2008) and participation by those with disabilities (Kelly, 2013).

Toronto's Pride Parade shows how varied subgroups bring other dimensions related to social justice into one celebratory event. In essence, identity goes beyond the narrow confines of sexual identity (Schnoor, 2006). Greensmith and Giwa (2013), in discussing Toronto's Pride Parade, also raise the importance of intersectionality regarding sexual identity, in particular by considering Indigenous culture and people.

Rosendahl (2012, 1) contextualizes Toronto's Pride Parade and introduces the role and importance of music, often a key element of celebratory events:

> Pride festivals are both political and celebratory in nature and often include parades, marches, street fairs, musical performances, and dance club events. Many view pride festivals as spaces used to create a unified group identity.

While such a view is partially accurate, the festival is also a space in which positions of centrality and marginality within the queer community are negotiated, particularly for gendered and racialized groups. With over one million participants and roughly 300 musical artists performing on multiple stages, Pride Toronto is one of the largest pride festivals in the world. As a leader on the global stage, Pride Toronto has struggled in recent years to create a festival that reflects the great diversity of the larger queer community in Toronto and abroad. The organization has been accused of marginalizing particular groups within the queer community through the uneven distribution of festival resources, the lack of organizational structures and advertisements aimed at particular sections of the queer community, and the placement of music stages and other areas directed at specific groups in undesirable locations and venues within the festival space ... Musical discourse, which includes not only the music and performance but also the programming and staging of musical artists, media reports, protests, and town hall meetings, was one of the primary means of initiating and sustaining dialogue on social power within the queer community in Toronto. Gendered and racialized groups used musical discourse to challenge power structures within the larger queer community, which had been highlighted by the allocation of time and space within the festival area.

Music and arts create an ambiance that is festive and helps attendees maintain attention over the course of a parade, which is no small achievement in situations where parades may necessitate hours of attention, and can take place during inclement weather (Bennett & Rogers, 2014). In addition to music, Toronto's Pride Parade has floats advertising local gay establishments (Shaw & Ardener, 2005) and has enjoyed major corporate sponsorships (Douglas, 2007).

Attendees at pride parades, in addition to enjoying the event, send a political message to the outside world that queer culture is an integral and constructive part of the social fabric of Canadian life (Ray, 2004). LGBTQ enclaves provide a political and social base from which members of this community can come together and use celebrations as a vehicle for intra-, inter-, and trans-community celebrations of urban culture.

Toronto's Pride celebration provides a window into how an event that started out as controversial over two decades ago has evolved and become an integral part of the Toronto landscape, illustrating the potential of parades to both attract positive reviews and accommodate controversies.

Conclusion

Toronto enjoys an international reputation, and as with any major city, its population represents great diversity. This diversity is manifested in myriad ways, including how the city celebrates its presence in neighborhoods and the city as a whole. There certainly is no denying the close association between parades and pride, be it ethnic, national, or LGBTQ, including the presence of intersectionality.

The Toronto parade case studies share commonality, yet also reflect national and cultural issues and sociopolitical priorities, highlighting how contextual forces influence parade manifestations and evolution as the groups sponsoring them increasingly feel more secure and look toward a future where they wield significantly more social, economic, and political influence. The staging of parades using urban public space and place as a backdrop to their struggles and aspirations illustrates the potential of events to channel energy and resources toward achieving these goals. Parades have long historical traditions throughout the world. Broadening our understanding that there is no one way of conducting a parade, parades will evolve and take on even greater prominence in a community's and nation's landscape.

SECTION 4

Reflections on Community Celebratory Events

11 Knowledge and Competencies for Community Practice and Celebratory Events

Although the role of practitioners has been discussed in this book, it has not been the focus of attention; yet practitioners can play a critical role in advancing the use of celebrations for achieving social and political goals. This chapter focuses on knowledge and skill-sets that are essential in order to move the field forward with a specific focus on urban-based celebratory events. These knowledge bases and skill-sets are not meant to be exhaustive but illustrative for tapping events to capture important moments in the life and experiences of communities.

From a community practice perspective, the knowledge base and skill-set for staging a celebratory event are of great importance. There are a number of outstanding books on event planning, as noted in the introductory chapter here. The authors of those books, however, focus on readers who have chosen to do event planning as a career, but that is not who *this* book is addressing. Opening the field to professionals other than event planners increases the possibilities of enhancing capacity and the benefits of an event to the community staging it.

Knowledge and competencies for engaging in community practice with regard to celebratory events are wide ranging (Getz, 2010, 2012; M.L. Jones, 2014). Community practitioners rarely have the option to specialize in one area, be it celebratory events or otherwise. Certainly it can be argued that events are more often staged by nonprofessionals than by professional event planners. Thus, events can become a part of the many different projects that community practitioners are involved in, and may even be seasonal.

Any opportunity to lift up a set of knowledge subjects and skill sets that are essential for a social intervention brings with it a set of challenges for a writer, and events are no exception. Does the author focus on identifying the knowledge areas and skill sets that can feasibly be acquired or should the focus be on acquiring those that are simply critical, regardless of feasibility of acquisition? Not surprisingly, this author has elected to address both, and hopes the reader will understand why this is the case.

Knowledge Areas

Celebratory events that have a specific focus on urban communities of minorities or those of newcomers, for example, necessitate drawing on knowledge areas that help us understand the sociocultural and sociopolitical context of the communities staging these events – going beyond the technical knowledge of how to stage a particular event (Bowdin et al., 2012). Skills are not possible without grounding in values and knowledge (Schwarz & Tait, 2007).

The following five subjects can easily be overlooked in the excitement and frantic pace of planning an event and many may be considered to be "soft" in significance: understanding the history of the community, understanding the cultural values and traditions of the groups sponsoring the events, working with local community media sources, identifying and enlisting community volunteers, and deciding how to record and save the history of the event.

Understanding the History of the Community

Appreciating the "context" in which an event transpires helps one to understand the forces at play in shaping how a celebration is conceptualized and unfolds. History is part of the context. History is often relegated to the fringes of the academy, and this conclusion is coming from an author who majored in history as an undergraduate and loved the subject.

An in-depth appreciation of a community's history is essential for planning, implementing, and evaluating events. More specifically, a historical grasp of the forces leading to newcomer groups entering specific neighborhoods, including the establishment of key social, economic, civic, and religious organizations, helps ground the importance of these institutions in the staging of the event (Ebert & Okamoto, 2013). This historical lens provides a guide for appreciating the meaning of celebratory events.

The community's history must be part of any discussion of context and practice. Urban histories of ethnic and racial groups are not easily obtained, particularly from the point of view of the community as history is most often written by society's elites. Practitioners may need to rely on oral history to obtain this historical perspective (Hansen, 2014). Depending on the subgroup, history and the interpretation of events varies according to the unique perch that each has (Nicholson, 2012). Youth, for example, view the significance of community and national events differently from their baby boomer and older adult counterparts.

Tying demographic data to newcomer influxes brings an added dimension to local history. Residents' stories illustrate how the demography of a community has changed, and these views are often missing in official documents reporting urban demographic changes. Familiarity with different demographic sources and how to analyze these data is a knowledge area that cannot be taken for granted. In cases where this content was not offered in formal schooling, practitioners will need to learn this on the job, which is probably not the easiest way of doing so.

Understanding the Cultural Values and Traditions of the Groups Sponsoring the Events

Appreciation of cultural values and traditions is associated with knowledge of a community's history (Delgado, 2007; Lee, Arcodia, & Lee, 2012). Such appreciation allows practitioners to set an event in cultural context, and this can supplement historical and demographic understandings of, for example, the forces that brought groups of immigrants to a new home and shaped their early and current existence (Derrett, 2003, 2004; Schwarz & Tait, 2007).

Understanding community assets or capital brings the potential of capacity enhancement to practice involving celebratory events. Cultural authenticity is not possible without a grounding in cultural values (Brown & James, 2004; Lytra, 2011). Cultural authenticity takes on significance with newcomer groups and their celebrations.

Culture is never static and must be broadly defined and operationalized. Youth have a culture beyond their ethnic or racial heritage, including language and concepts that are age-specific. Further, acculturation places different members of families at different points in a continuum of values and traditions (White, 2012). Where a family is new in an English-speaking homeland, the older family members will be more likely to return to traditional values and less likely to adopt English as their primary language; youth, in turn, are more likely to modify or even discard traditional values in favor of the more mainstream values of their country. For younger family members, English is more likely to be their primary language compared with their parents.

Community practitioners must traverse levels of cultural acculturation to increase their effectiveness in reaching different subgroups and encouraging them to participate in staging a celebratory event (Reverte & Izard, 2011). This is a tall order. Nevertheless, this is a goal worth achieving in order to make community celebrations an integral part of urban practice.

Working with Local Media Community Sources

Working with local community media can be considered both a knowledge area and highly specialist content and skill-set (Wardle & West, 2004). Local media such as cable TV shows, radio programs, and community newsletters are important sources for getting out information to community groups (Couser et al., 2007; McCabe, 2006). Hood (2007), however, issues a cautionary note on news reporting trends among local radio stations.

Local urban media, particularly print and radio, take on even greater importance when the targets groups do not have English as their primary language – because conventional English media sources simply do not penetrate these communities and therefore are irrelevant in these communities. Organizational and religious institutional newsletters, too, must not be overlooked. Social media outlets, in turn, have particular appeal to some population subgroups and take on significance in reaching out to them in a manner that gets their attention and is message-sensitive (Gerodimos, 2010).

Identifying and Enlisting Community Volunteers (Civic Engagement)

Civic engagement can be considered a key element in the ownership of community events (Carson, Chappell, & Knight, 2007; Sumner, Mair, & Nelson, 2010). Knowing this enhances community practice and capacity in the process (Monga, 2006). Civic engagement is salient in community practice and the staging of celebratory events, thus volunteering is an important matter (Arcodia & Whitford, 2006; Procter, 2004; Ziakas & Costa, 2012).

Community participation is essential for events to be owned by the community, and community participation shapes the goals and messages these events convey internally and externally. Successful volunteer programs do not simply appear without serious attention and resources being directed to them by the event organizers (Johnston, Twynam, & Farrell, 1999). Possessing a deep and profound grasp of how civic engagement can enhance celebratory events will be indispensable for practitioners (Giannoulakis, Wang, & Gray, 2007).

Special attention must be paid to having a broad segment of the community involved in staging events and to not excluding groups due to biases, by scheduling activities during time periods that do not facilitate a subgroup's involvement (e.g., school hours and youth), or by recruiting participants in circles that refuse to come into contact with certain groups. Bias may be conscious or unconscious, and practitioners can point out bias in the planning and staging of events.

Deciding How to Record and Save the History of the Event

Archiving event knowledge performs an important service to communities (Bowdin et al., 2012). Collective experiences translate into collective memories, and recording these memories helps communities understand and embrace their past. The critical role in documenting, digital and/or print), an event becomes part of a community's collective memory and history, helping to preserve an important part of a community's lived experience (Williams & Culp, 2011; Zigkolis et al., 2014).

Capturing celebratory events for future generations has been addressed (Stockinger, 2013). Involving the local residents in this endeavor increases the community's capacity and allows other communities to learn from each other's experiences (Bellinger, 2013). Involving youth, particularly with their social media/digital knowledge and the fact that they will potentially be a part of the community for many decades to come, becomes very attractive for practitioners and the community. The storage of these archives in easily accessed places should follow the recording of celebratory events (Silverman, 2010). Practitioners are not expected to archive the celebratory event data, but they must be aware of the importance of doing so and the ways that this history can be recorded and retrieved. Serving to broker resources that specialize in archiving information can be a role for practitioners, and enlisting local librarians and university resources, for example, helps in achieving this goal.

Competencies Needed for Community Practice

Celebratory events are not the efforts of one person and are best conceptualized as collective community efforts. Can any community practitioner conduct a celebratory event and incorporate it into her or his practice? No, but that is not unique to celebratory events and applies to all other forms of community practice. Some skill-sets were given attention throughout earlier portions of this book; others may not appear as clearcut, but are of no less importance than those that are obvious (Clifton, O'Sullivan, & Pickernell, 2012).

The subject of event planner competencies has received considerable attention in the professional literature (Raj, Walters, & Rashid, 2013). The following competencies stand out for in-depth attention in this section: the use of the media, language skills, group skills, and the development of a community of practitioners.

The professional literature has taken a more broad-based approach to event practice. Further, this approach has not emphasized geographical setting, an essential contextual factor in shaping how events unfold and how they are viewed by sponsoring communities. The transferring of these skills to community residents and event leaders represents an important dimension that is emphasized when using a capacity-enhancement paradigm, but has not been emphasized enough in the events literature (Ziakas & Costa, 2011).

Use of Media

Community practice in any form must integrate the effective use of media. Media coverage of celebratory events takes on greater prominence when employing a community capacity enhancement paradigm. Why should media play such an important role in this paradigm? Capacity enhancement involves changing the physical environment, mobilizing external and internal resources, and enlisting residents in achieving this goal through projects such as gardens, murals, sculptures, and playgrounds, and in the case of this book, celebratory events. The publicizing of these changes takes on great significance when employing this paradigm.

Media coverage, when positive, serves to reinforce a positive image of the community within and outside the community. Incidentally, the use of large screens opens up the possibility of having celebratory events reach out beyond the geographical confines of where they are held to include a broader audience (Mcquire, 2010). In addition, media coverage (visual or print) of a celebratory event provides a record, serving as a recorded memory that a community can share in the future.

The emergence of social media as a key force in democratizing information has put this power within the hands of ordinary community residents (Ang, 2011; Getz, 2012) and has shifted conventional media relations strategies (Bowdin et al., 2012; Waters, Tindall, & Morton, 2010). This movement has literally placed the recording of history into every resident's hands, together with the power that it brings for shaping the interpretation and meaning of events.

Language Skills

Those in the field of urban event planning argue that language competency beyond English is as an indispensable tool for engaging newcomer urban groups (Geiger et al., 2013). This is certainly the case in particular regions of the United States and Canada. Language also takes on a nuanced perspective because community practitioners must be capable of speaking to individuals with many different formal educational backgrounds, including different age groups. Being able to engage in discourse in a manner that is not condescending is not to be minimized (Abrahams, 2012).

It would be ideal for a practitioner to be bilingual (Hutchins, Brown, & Poulsen, 2014). Even making a serious effort to develop a working vocabulary will go a long way toward creating good will and a relationship that can grow over time. Few practitioners realize that communities composed of non-English speakers rarely encounter English-speaking people willing to even make an effort to speak a few words, making efforts at speaking another language special and noteworthy (Walker & Polepeddi, 2013).

Group Skills

Interpersonal skills are at the crux of any form of community practice. The professional literature on event planning has overlooked the importance of group skills in the planning, implementation, and evaluation of community celebratory events. It would be a rare higher educational program where a course on group skills would be offered as an elective yet alone a required course. Practitioners must acquire these skills in the field, making it more challenging than it has to be.

So much happens within groups such as advisory committees, stakeholder groups, work groups, and meetings of various kinds – not to mention the groups that attend celebratory events. Group skills take on greater significance in work with communities, and no practitioner can eschew working with groups, yet not every practitioner has group skills.

Skills in working with groups cover a range from ensuring the composition of committees is inclusive to having task-oriented skills that help committees process information and make decisions. Helping to recruit, foster, and support community group leaders is an important function for community practitioners to perform, and this support transpires in individual sessions and in groups, too.

Development of a Community of Practitioners

The evolution of event planning as a profession is undeniable (Crowther, 2010). Dickson and Arcodia (2010) specifically address the importance of professional organizations in promoting sustainable event practice. However, knowledge generated in these associations represents but one aspect of event planning (Arcodia & Reid, 2003). A community of practitioners ensures that event information and support are available. Being able to assist in the development of a community of practitioners is not the same as being the leader and chief developer of this association.

A facilitative role brings together practitioners and academics to discuss community celebrations as a field of practice, resulting in rewards for practitioners and communities alike. As this field grows, practitioners will share their experiences, documents, and contacts, making practice that much easier, making the investment of time and energy worthwhile.

Summary

The competencies identified can be quite intimidating. No practitioner possesses all of these competencies. Nevertheless, these competencies highlight the complexity of community practice, and cannot be the solely the responsibility of practitioners obtaining them in the field post-graduation. Opportunities to develop and enhance knowledge and competencies must be present in graduate community practitioner programs.

This chapter provides recommendations for moving this field forward. These recommendations encompass research, practice, and education, in order to provide a comprehensive perspective on the use of celebratory events to enhance local capacities. The tasks covered in this chapter can realistically be accomplished, although none of them is to be considered "easy" to master in practice or academia.

Nevertheless, urban communities must acknowledge the importance of the field of celebratory events being open to community practitioners. Academia must be prepared to make changes to the "legitimacy" of fostering events. Urban events will continue to occur without these two worlds coming together – until they do so, however, the true potential of celebratory events will elude communities.

Concluding Thoughts

There is always a need to have space devoted to content that either was not addressed in an in-depth manner or simply could not find the right space. These concluding thoughts provide the "place" and "space" to share parting thoughts, not encumbered by relying on scholarly definitions, quotations, or citations, but an opportunity to reflect with an understanding that the questions posed cannot be answered at this point in time, but are worth noting nonetheless.

Are Community Celebratory Events Needed in All Urban Communities?

All communities can benefit from a collective experience when residents come together to share a moment of joy. The resulting collective memory can cross generations. Urban communities are in need of opportunities when they can gather to celebrate because there are such few moments when they can do so. Celebratory events are not to be forced on communities as if "you must celebrate now or else."

Marginalized communities face incredible odds for surviving the destructive global economic and social trends associated with increasing income disparities and lack of upward mobility. Being trapped can destroy dreams and relegate generations to the margins of society, and making those experiencing these challenges feel they are alone in these struggles.

Will community celebratory events counter these mega-trends? I wish I could say yes. However, the answer is "no," but this does not take away from the potential of celebratory events to transform communities because such events affirm and bring together disparate groups. Postindustrialized societies are becoming more age-segregated, with few opportunities for an entire family to attend an event, and particularly an event that is either free or necessitates minimal cost outlays, and to do so with a sense of freedom that comes with a "safe" event. Celebratory events are the exceptions, bridging differences within and outside communities, and within a budget and timeframe.

Corporations to the Rescue?

Events often require external funding although much can be accomplished through civic engagement and minimizing reliance on external funding. There is no denying the attractiveness of external funding. The increasing role that corporations are playing in sponsoring celebratory events must be signaled out, and their influence will continue to increase in the immediate future as corporations find events attractive for targeting highly segmented and difficult-to-reach groups.

These consequences must be checked if community practice as it involves celebratory events is not to be subverted. The infusion of external money to pay for celebratory expenses is attractive from a staging perspective. However, money is never given without strings. If corporate funding is necessary, every effort must be made to solicit sponsorships from community-owned businesses rather than large corporate sponsors based outside of the hosting community.

Local businesses are attractive because of their direct history and investment in their local markets and communities, and they are less likely to extract political capital for providing financial support (Delgado, 2012). Owners of local businesses often reside within the communities they serve, increasing their interests in local matters. When financial support is sought from local businesses it is important that these businesses have the support of the community as a check and balance on the exercise of power. Local money translates into local control and power.

Evaluation, Evaluation, Evaluation, and Further Evaluation of Celebratory Events

The response to this topic is especially long but necessary because of the role and importance of this phase of any community celebratory event. The subject of evaluation is destined to take on greater importance in the immediate future, necessitating that academics and community practitioners broach this topic in a systematic and collaborative manner, with communities playing an active and decision-making role.

Evaluation necessitates the quest for innovative methods that are user-friendly but also capture nuances of celebratory events that are extremely important but easily overlooked when using more conventional methods. Community events are very context driven, and dependent on a keen understanding of how local forces influence the unfolding of these events.

No evaluation method can do justice to the wide impact of an urban celebratory event on the community's capital, or assets, particularly where a highly diverse community is the backdrop, and with principles of participatory democracy guiding these events. How is *community* defined, and do all vested (stakeholders) parties agree on this definition? These are critical questions that may defy a consensus of opinion. Do all participants in the event gain equally from their work or do some subgroups benefit more or less than others? This last question takes on greater importance when there are efforts to include groups that historically have not been part of community activities.

Evaluations require both qualitative and quantitative methods, including the development of new methods or the use of methods not normally associated with evaluating community celebratory events. Arts-based inquiry offers great promise for use in impact analysis because of how it taps creative approaches that can be culture-specific, participatory, and empowering in the process (Delgado, 2015). Arts-based inquiry brings excitement to the evaluation process and challenges in integrating this approach with conventional methodology.

Ethnography takes on increased relevance where community celebratory events are deeply rooted in cultural heritage and emphasize local history and circumstances, making it useful in evaluation (Gioia, 2014; Madison, 2011; Stadler, Reid, & Fullagar, 2013). Holloway, Brown, and Shipway (2010, 74) describe the potential of ethnography and celebratory events: "Events research is witnessing a gradual increase in experience-related studies, reflecting a challenge to the dominance of positivist, quantitative-based studies ... The literature on qualitative methods and on ethnography in particular ... show how ethnography can be used and how it is specifically suited to inquiry into the consumer experience of events and festivals. Ethnography is advocated as an appropriate research approach to the events field."

Biaett's (2012) unconventional approach toward evaluating festivals uses an auto-ethnographic perspective as a response to calls for innovative methods that can capture a contextual understanding of festival attendance from the attendees' point of view during, and after, attendance. Crooks (2013), too, advocates for this approach with the LGBTQ community.

This form of epistemology, research design, and measurement method brings challenges for capturing attendees' experiences in a manner that lays the groundwork for a spirited debate, advancing the field of community celebratory events. The search for knowledge must take into account the unit of analysis. Is the focus of events on individuals or on the collective? If the latter, it introduces an important, but no less challenging view of how these events transform communities and not just individuals.

Jaimangal-Jones (2014, 39) addresses the challenges associated with using an ethnographic approach to evaluating celebratory events: "The challenges facing ethnographic researchers studying event audiences include identifying opportunities for observation and participation, identity negotiation for different research settings, their positioning on the participant observer spectrum, recruiting participants, recording data and the extent to which research takes an overt or covert approach, bearing in mind ethics and participant reactivity." Evaluating without interfering is a worthy goal for all evaluators of community celebratory events. Nowhere is this more apparent than with using participatory qualitative methods. Critical ethnography provides an approach to and contextual grounding of those most marginalized (Madison, 2011).

Critical epistemology can be expected to take on increased significance in the field of celebratory events (Biaett, 2012). Use of this and other methods is a tall order for any evaluator but essential for this field to progress. In addition, critical epistemology helps shine light on subgroups that are further marginalized because of their multiple jeopardies.

There is a desperate need for multidisciplinary team approaches to community celebratory events. This recommendation always looks good on paper. Yet it means bringing together professions that may not share similar paradigms, language, research methods, and possibly have histories of mistrust, if not antagonism. Much trust-focused work needs to be undertaken before people from different disciplines can work together as an "evaluation team."

Involving the community's residents increases collaboration challenges. Planning community events must take into account people with a wide range of formal educational levels. A team approach brings a needed perspective to the evaluation of celebratory events that has been missing in the past. Community participation in the evaluation of an event helps evaluators acquire insights and make recommendations that are more meaningful for the community and the funders.

Does Type of Event and Size Matter?

This is a very important question. My answer may appear simplistic. Nevertheless, the answer is "it depends." In some highly marginalized communities, the type and size of a community celebratory event does not matter because the actual staging represents that important initial step in a community seeking to take stock of its heritage and assets and wishing to make a positive statement about itself to the outside world. Nevertheless, small events rarely bring forth a "big bang" or widespread publicity, have sizable evaluation budgets, or even pay attention to evaluation.

The goals of a community celebratory event and the political will of the sponsoring groups dictate the size and elaborateness of the event. A community with a limited history of sponsoring celebratory events is not in a position to undertake a highly elaborate event or evaluation, while communities with long histories of launching events are in a propitious position to undertake large and multifaceted events and command funding for an extensive evaluation. Local historical context shapes the answer to the question of whether size matters in a celebratory event.

The proliferation of community celebratory events makes sponsoring these events more common and less special. Careful thought has to be given to what types of events, how often a community should stage them, and their size and complexity if the events are to wield significance in the life of a community. Celebratory events must be viewed from a strategic perspective because of the time, energy, and resources needed to ensure their success. A zero sum approach must be taken to event sponsoring because other community projects must then be set aside.

Is This Form of Community Practice for All Community Practitioners?

This question is important because community social work practice covers such a wide territory, and it would be unreasonable to expect practitioners to embrace and be equally competent in all arenas and facets associated with this form of practice. The competencies and qualities outlined in chapter 11 apply in answering this question. Further, the answer has ramifications for educational programs, requiring a close relationship between the event management industry and academia (Jiang & Schmader, 2014).

This question is best addressed with an understanding that community celebratory events have various phases and specific tasks related to each phase. No community practitioner can, or should, be expected to be highly competent with all of these facets. Being able to specialize in particular phases and tasks and seek assistance when it is needed are marks of a professional.

Is There a Particular Type of Community Organization that Should Spearhead Community Celebratory Events?

The answer to this question is tied to who should undertake events. Can any organization play a leading or significant role in staging a celebratory event? The answer is clearly "no!" Community celebratory events are special and that necessitates that organizations sponsoring and staging them be special, too. Organizations that have a history of participatory democracy and collaborating with community institutions, both formal and nontraditional, are in a propitious position to play a leading role. Nontraditional organizations are places where residents go to purchase a product or service, or to congregate, and in the process receive social and health services (Delgado, 1999a).

There are few community organizations that devote their mission solely to staging celebratory events. Organizations sponsoring or staging community events generally do so as an added-on function. A determination of which community organization should be the leading one in putting on a celebratory event depends on the following factors: history of collaborative undertakings and a willingness to participate in partnerships; legitimacy within the community; personnel with the requisite competencies; its positive relationship with local governmental agencies; the financial capabilities to underwrite and obtain external funding; and a keen understanding of the competencies needed to handle the public relations functions associated with celebratory events. Ideally, all seven factors should be satisfied before a community organization takes the lead in staging an event.

What Is the Future of Community Celebratory Events?

It is appropriate to end this chapter, section, and book with the question of what the future holds in store for community celebratory events. Future ethnic celebrations, for example, must contend with commercialization as a key staging motivator, as noted by Hayes-Bautista (2012, 191): "It is interesting to speculate about what form future celebrations of the holiday [Cinco de Mayo] might take, should its true origins and heritage become better understood. Naturally, the blatantly commercial aspects will not disappear; by now, virtually no American holiday has escaped some degree of commercialization." Commercialization is a perennial theme, raising the specter of how money and power can corrupt.

Celebrations tap human potential even when communities are facing incredible challenges, tapping into assets. Acknowledging that a community has reason to celebrate with a time and place for coming together is critical. What better way than through a collective and affirming effort that brings all segments of the community together in pursuit of a common goal, in a highly public manner, with potential for making a collective memory and spreading learning about the community?

Conclusion

This final chapter has addressed some obvious and not-so-obvious themes, including some that are quite complex and challenging for community social work practice. Simplifying a world view of community celebratory events within this context of practice would be a disservice to the field because these events are complex. When participatory principles are introduced, event organization and evaluation becomes messy because democracy is messy. Although this book's focus has been on celebratory events in urban communities, they certainly are not restricted to this setting. The future is bright for community celebratory events, particularly when viewed strategically. Community practitioners are in an excellent position to serve as the "glue" for bringing together community residents and professionals in this field.

References

Abdulla, N. (2013). *Stakeholder Perspectives on How Tourism Development Is Undertaken in Waterloo Region*. Doctoral dissertation, University of Waterloo, Waterloo, ON.

Abebe, N.A., Capozza, K.L., Des Jardins, T.R., Kulick, D.A., Rein, A.L., Schachter, A.A., & Turske, S.A. (2013). Considerations for community-based mHealth initiatives: Insights from three Beacon Communities. *Journal of Medical Internet Research, 15*(10), e221. http://dx.doi.org/10.2196/jmir.2803

Abidin, I.Z., Usman, I.M., & Tahir, M.M. (2010). Characteristic of attractive square as public space: Putra Square, Putrajaya. In M. Mazilu & J. Strouhal (Eds.), *Selected Topics in Energy, Environment, Sustainable Development and Landscaping*, 338–43. Sofia, Bulgaria: WSEAS.

Abrahams, R.D. (2012). Questions of competency and performance in the black musical diaspora: Toward a stylistic analysis of the idea of a black Atlantic. *Black Music Research Journal, 32*(2), 83–93. http://dx.doi.org/10.5406/blacmusiresej.32.2.0083

Abu, N. (2012). *Consumer Decision-Making on Festival Attendance*. Business Management Bachelor's thesis, Laurea University of Applied Sciences, Espoo, Finland. http:urn.fi/URN:NBN:fi:amk-201204024012

Abulafia, A., Segev, F., Platner, E., & Ben Simon, G.J. (2013). Party foam-induced eye injuries and the power of media intervention. *Cornea, 32*(6), 826–9. http://dx.doi.org/10.1097/ICO.0b013e31826cf315

Ackermann, O., Lahm, A., Pfohl, M., Köther, B., Lian, T.K., Kutzer, A., ..., & Hax, P.M. (2011). Patient care at the 2010 Love Parade in Duisburg, Germany. *Deutsches Ärzteblatt International, 108*(28–29), 483–9.

Adams, J.D. (2013). Theorizing a sense of place in a transnational community. *Children, Youth and Environments, 23*(3), 43–65. http://dx.doi.org/10.7721/chilyoutenvi.23.3.0043

Adams, N.L., & Adams, L.L. (2012). Gravitational forces of community events: A new method of evaluating value of local events. *American Journal of Management*, *12*(1), 131–48.

Alem, S., Vaziri, V., & Sharif, A.R. (2014). Designing ritual spaces by considering social bonds. *Journal of Social Issues & Humanities*, *2*(5), 12–19.

Alivizatou, M. (2012). Debating heritage authenticity: *Kastom* and development at the Vanuatu Cultural Centre. *International Journal of Heritage Studies*, *18*(2), 124–43. http://dx.doi.org/10.1080/13527258.2011.602981

Al-Kodmany, K. (2013). Crowd management and urban design: New scientific approaches. *URBAN DESIGN International*, *18*(4), 282–95. http://dx.doi.org/10.1057/udi.2013.7

Allen, B.A., EchoHawk, D., Gonzales, R., Montoya, F.-A., & Somerville, M.M. (2012). *Yo Soy* Colorado: Three collaborative Hispanic cultural heritage initiatives. *Collaborative Librarianship*, *4*(2), 39–52.

Alonso, A.D., & O'Shea, M. (2012). "You only get back what you put in": Perceptions of professional sport organizations as community anchors. *Community Development*, *43*(5), 656–76. http://dx.doi.org/10.1080/15575330.2011.645048

Al-Tarshihi, M.I., & Al-Basheer, M. (2014). The falling bullets: Post-Libyan revolution celebratory stray bullet injuries. *European Journal of Trauma and Emergency Surgery*, *40*(1), 83–5. http://dx.doi.org/10.1007/s00068-013-0323-1

Altschuld, J.W. (2015). *Bridging the Gap between Asset/Capacity Building and Needs Assessment: Concepts and Practical Applications*. Los Angeles: Sage.

American Planning Association. (2011). *How Arts and Cultural Strategies Create, Reinforce, and Enhance Sense of Place*. http://www.planning.org/research/arts/briefingpapers/character.htm

Amerson, R. (2010). The impact of service-learning on cultural competence. *Nursing Education Perspectives*, *31*(1), 18–22.

Amin, A. (2008). Collective culture and urban public space. *City: Analysis of Urban Trends, Culture, Theory, Policy, Action*, *12*(1), 5–24. http://dx.doi.org/10.1080/13604810801933495

Amundsen, H. (2012). Illusions of resilience? An analysis of community responses to change in Northern Norway. *Ecology and Society*, *17*(4), 357–70. http://dx.doi.org/10.5751/ES-05142-170446

Anderson, K.J. (2014). The idea of Chinatown: The power of place and institutional practice in the making of a racial category. In J.J. Gieseking & W. Mangold (Eds.), *The People, Place, and Space Reader*, 87–91. New York: Routledge. http://dx.doi.org/10.1111/j.1467-8306.1987.tb00182.x

Andersson, T.D., Armbrecht, J., & Lundberg, E. (2012). Estimating use and non-use values of a music festival. *Scandinavian Journal of Hospitality and Tourism*, *12*(3), 215–31. http://dx.doi.org/10.1080/15022250.2012.725276

Andersson, T.D., & Getz, D. (2008). Stakeholder management strategies of festivals. *Journal of Convention & Event Tourism, 9*(3), 199–220. http://dx.doi.org/10.1080/15470140802323801

Andersson, T.D., Getz, D., & Mykletun, R. (2013a). The "festival size pyramid" in three Norwegian festival populations. *Journal of Convention & Event Tourism, 14*(2), 81–103. http://dx.doi.org/10.1080/15470148.2013.782258

Andersson, T.D., Getz, D., & Mykletun, R.J. (2013b). Sustainable festival populations: An application of organizational ecology. *Tourism Analysis, 18*(6), 621–34. http://dx.doi.org/10.3727/108354213X13824558188505

Andersson, T.D., & Lundberg, E. (2013). Commensurability and sustainability: Triple impact assessments of a tourism event. *Tourism Management, 37*(Aug.), 99–109. http://dx.doi.org/10.1016/j.tourman.2012.12.015

Anderton, C. (2011). Music festival sponsorship: Between commerce and carnival. *Arts Marketing: An International Journal, 1*(2), 145–58. http://dx.doi.org/10.1108/20442081111180368

Andriotis, K. (2011). Genres of authenticity: Denotations from a pilgrimage landscape. *Annals of Tourism Research, 38*(4), 1613–33. http://dx.doi.org/10.1016/j.annals.2011.03.001

Ang, L. (2011). Community relationship management and social media. *Journal of Database Marketing & Customer Strategy Management, 18*(1), 31–8. http://dx.doi.org/10.1057/dbm.2011.3

Anguelovski, I. (2014). *Neighborhood as Refuge: Community Reconstruction, Place Remaking, and Environmental Justice in the City*. Cambridge, MA: MIT Press. http://dx.doi.org/10.7551/mitpress/9780262026925.001.0001

Ansari, S., Munir, K., & Gregg, T. (2012). Impact at the "bottom of the pyramid": The role of social capital in capacity development and community empowerment. *Journal of Management Studies, 49*(4), 813–42. http://dx.doi.org/10.1111/j.1467-6486.2012.01042.x

Antony, M.G. (2014). "Hello, how may I offend you today?": NBC's Outsourced and the discourse of cultural authenticity. *Communication Review, 17*(1), 1–26. http://dx.doi.org/10.1080/10714421.2014.872497

Araoz, G.F. (2011). Preserving heritage places under a new paradigm. *Journal of Cultural Heritage Management and Sustainable Development, 1*(1), 55–60. http://dx.doi.org/10.1108/20441261111129933

Arcodia, C.V., & Reid, S.L. (2003). Professionalising event practitioners: The educational role of event management associations. In K. Weber (Ed.), *Advances in Convention, Exhibition and Event Research, Convention and Expo Summit*, 29–31. Hong Kong: Hong Kong Polytechnic University.

Arcodia, C., & Whitford, M. (2006). Festival attendance and the development of social capital. *Journal of Convention & Event Tourism, 8*(2), 1–18. http://dx.doi.org/10.1300/J452v08n02_01

Arnoldi, M.J. (2006). Youth festivals and museums: The cultural politics of public memory in postcolonial Mali. *Africa Today, 52*(4), 55–76. http://dx.doi.org/10.1353/at.2006.0037

Aron, J. (2011). Smart software spots a dangerous crowd. *New Scientist, 211*(2824), 23. http://dx.doi.org/10.1016/S0262-4079(11)61890-3

Arora, P. (2015). Usurping public leisure space for protest: Social activism in the digital and material commons. *Space and Culture, 18*(1), 55–68.

Arroyo-Martinez, J. (2012). Review of *Cuban fiestas* by Roberto González Echevarría. *Review: Literature and Arts of the Americas, 45*(1), 138–9. http://dx.doi.org/10.1080/08905762.2012.670502

Ashebir, W.W.K. (1999). *Urban Rituals: A Study of Religious and Secular Rituals as an Informer of the Built Environment*. Doctoral dissertation, University of the Witwatersrand, South Africa.

Ashworth, G., & Page, S.J. (2011). Urban tourism research: Recent progress and current paradoxes. *Tourism Management, 32*(1), 1–15. http://dx.doi.org/10.1016/j.tourman.2010.02.002

Aslimoski, P., & Gerasimoski, S. (2012). Food and nutrition as tourist phenomenon. *Procedia: Social and Behavioral Sciences, 44*, 357–62. http://dx.doi.org/10.1016/j.sbspro.2012.05.039

Ayob, N., Wahid, N.A., & Omar, A. (2013). Mediating effect of visitors' event experiences in relation to event features and post-consumption behaviors. *Journal of Convention & Event Tourism, 14*(3), 177–92. http://dx.doi.org/10.1080/15470148.2013.814037

Azar, K.M.J., Chen, E., Holland, A.T., & Palaniappan, L.P. (2013). Festival foods in the immigrant diet. *Journal of Immigrant and Minority Health, 15*(5), 953–60. http://dx.doi.org/10.1007/s10903-012-9705-4

Bada, X. (2014). *Mexican Hometown Associations in Chicagoacán: From Local to Transnational Civic Engagement*. New Brunswick, NJ: Rutgers University Press.

Bagiran, D., & Kurgun, H. (2013). A research on social impacts of the Foça Rock Festival: The validity of the Festival Social Impact Attitude Scale. *Current Issues in Tourism*. http://dx.doi.org/10.1080/13683500.2013.800028

Bagri, N.T. (2014, 24 Aug.). India's towering human pyramids draw crowds and fears. *New York Times*, 7.

Bairner, A. (2011). Urban walking and the pedagogies of the street. *Sport Education and Society, 16*(3), 371–84. http://dx.doi.org/10.1080/13573322.2011.565968

Bakshi, A. (2014). Urban form and memory discourses: Spatial practices in contested cities. *Journal of Urban Design, 19*(2), 189–210. http://dx.doi.org/10.1080/13574809.2013.854696

Balaceanu, C., Apostol, D., & Penu, D. (2012). Sustainability and social justice. *Procedia: Social and Behavioral Sciences, 62*, 677–81. http://dx.doi.org/10.1016/j.sbspro.2012.09.115

Ball, W. J., & Wanitshka, C. (2014). Green fairs as venues for civic engagement. *Local Environment: The International Journal of Justice and Sustainability*. http://dx.doi.org/10.1080/13549839.2014.914900

Ballester, P. (2014). Leisure parks from the Spanish universal and international exhibitions. *Loisir et Société/Society and Leisure, 37*(1), 38–57. http://dx.doi.org/10.1080/07053436.2014.881091

Balomenou, N., & Garrod, B. (2014). Using volunteer-employed photography to inform tourism planning decisions: A study of St David's Peninsula, Wales. *Tourism Management, 44*(2), 126–39. http://dx.doi.org/10.1016/j.tourman.2014.02.015

Banerjee, T., Uhm, J., & Bahl, D. (2014). Walking to school: The experience of children in inner city Los Angeles and implications for policy. *Journal of Planning Education and Research, 34*(2), 123–40. http://dx.doi.org/10.1177/0739456X14522494

Bannister, J., & Kearns, J. (2013). The function and foundations of urban tolerance: Encountering and engaging with differences in the city. *Urban Studies, 50*(13), 2700–17.

Barber, N.A., Kim, Y.H., & Barth, S. (2014). The importance of recycling to U.S. festival visitors: A preliminary study. *Journal of Hospitality Marketing & Management, 23*(6), 601–25.

Barker, E., O'Gorman, J., & de Leo, D. (2014). Suicide around public holidays. *Australasian Psychiatry, 22*(2), 122–6. http://dx.doi.org/10.1177/1039856213519293

Barnes, C. (2003). Effecting change: Disability, culture and art? Paper presented at the Finding the Spotlight Conference, Liverpool Institute for the Performing Arts, 28–31 May, Liverpool.

Baron-Yelles, N., & Clave, S.A. (2014). Leisure parks: Components and creators of the new urban landscapes? *Loisir et Société/Society and Leisure, 37*(1), 18–37. http://dx.doi.org/10.1080/07053436.2014.881090

Barrientos, J., Silva, J., Catalan, S., Gómez, F., & Longueria, F. (2010). Discrimination and victimization: Parade for lesbian, gay, bisexual, and transgender (LGBT) pride, in Chile. *Journal of Homosexuality, 57*(6), 760–75. http://dx.doi.org/10.1080/00918369.2010.485880

del Barrio, M.J., Devesa, M., & Herrero, L.C. (2012). Evaluating intangible cultural heritage: The case of cultural festivals. *City, Culture and Society*, *3*(4), 235–44.

Barrios, R.E. (2010). You found us doing this, this is our way: Criminalizing Second Lines, Super Sunday, and habitus in post-Katrina New Orleans. *Identities: Global Studies in Culture and Power*, *17*(6), 586–612. http://dx.doi.org/10.1080/1070289X.2010.533522

Barron, J. (2011, 2 May). A metal-gate makeover. *New York Times*, 35.

Barron, P., & Rihova, I. (2011). Motivation to volunteer: A case study of the Edinburgh International Magic Festival. *International Journal of Event and Festival Management*, *2*(3), 202–17. http://dx.doi.org/10.1108/17582951111170281

Barthel, S., Parker, J., & Ernstson, H. (2015). Food and green space in cities: A resilience lens on gardens and urban environmental movements. *Urban Studies*, *52*(7), 1321–38.

Barthel, S., Parker, J., Folke, C., & Colding, J. (2014). Urban gardens: Pockets of social-ecological memory. In K.G. Tidball & M.E. Krasny (Eds.), *Greening in the Red Zone: Disaster, Resilience and Community Greening*, 145–58. Heidelberg: Springer.

Bartoş, S.E., Balş, M.A., & Berger, I. (2014). Since Trajan and Decebalus: Online media reporting of the 2010 GayFest in Bucharest. *Psychology & Sexuality*, *5*(3), 268–82.

Barua, M. (2014). Circulating elephants: Unpacking the geographies of a cosmopolitan animal. *Transactions of the Institute of British Geographers*, *39*(4), 559–73. http://dx.doi.org/10.1111/tran.12047

Barzan, A., Bonne, B., Quax, P., Lamotte, W., Versichele, M., & Van de Weghe, N. (2013). A comparative simulation of opportunistic routing protocols using realistic mobility data obtained from mass events. Presented at IEEE 14th International Symposium and Workshops on World of Wireless, Mobile and Multimedia Networks (WoWMoM), 4–7 June, Madrid.

Bastian, J.A. (2013). The records of memory, the archives of identity: Celebrations, texts, and archival sensibilities. *Archival Science*, *13*(2–3), 121–31. http://dx.doi.org/10.1007/s10502-012-9184-3

Bates, R.A., & Fortner, E.B. (2013). The social construction of frivolity. *Journal of Public and Professional Sociology*, *5*(1), art5.

Bathelt, H., & Spigel, B. (2012). The spatial economy of North American trade fairs. *Canadian Geographer/Geographe Canadien*, *56*(1), 18–38. http://dx.doi.org/10.1111/j.1541-0064.2011.00396.x

Batty, M., Desyllas, J., & Duxbury, E. (2003a). The discrete dynamics of small-scale spatial events: Agent-based models of mobility in carnivals and street parades. *International Journal of Geographical Information Science, 17*(7), 673–97. http://dx.doi.org/10.1080/1365881031000135474

Batty, M., Desyllas, J., & Duxbury, E. (2003b). Safety in numbers? Modelling crowds and designing control for the Notting Hill Carnival. *Urban Studies, 40*(8), 1573–90. http://dx.doi.org/10.1080/0042098032000094432

Baum, T., Lockstone-Binney, L., & Robertson, M. (2013). Event studies: Finding fool's gold at the rainbow's end. *Event Management, 4*(3), 179–85.

Baycan, A., & Nijkamp, P. (2012). Critical success factors in planning and management of urban green spaces in Europe. *International Journal of Sustainable Society, 4*(3), 209–25. http://dx.doi.org/10.1504/IJSSOC.2012.047278

Beard, V.A., & Sarmiento, C.S. (2010). Ties that bind: Transnational community-based planning in Southern California and Oaxaca. *International Development Planning Review, 32*(3–4), 207–24. http://dx.doi.org/10.3828/idpr.2010.06

Becker, K. (2013). Performing the news. *Photographies, 6*(1), 17–28. http://dx.doi.org/10.1080/17540763.2013.788833

Beckley, T.M., Martz, D., Nadeau, S., Wall, E., & Reimer, B. (2008). Multiple capacities, multiple outcomes: Delving deeper into the meaning of community capacity. *Journal of Rural and Community Development, 3*(1), 56–75.

Beider, H. (2011). *Community Cohesion: The Views of White Working-Class Communities*. London: Joseph Rowntree Foundation.

Belghazi, T. (2006). Festivalization of urban space in Morocco. *Critique, 15*(1), 97–107. http://dx.doi.org/10.1080/10669920500515168

Bell, C. (2014). Cultural practices, market disorganization, and urban regeneration: Royal Court Theatre Local Peckham and Peckham space. *Contemporary Theatre Review, 24*(2), 192–208. http://dx.doi.org/10.1080/10486801.2014.885897

Bell, J., Radford, K., & Young, O. (2013). *Community Dialogue Tool: The Lurgan Town Project*. Belfast: Institute of Conflict Research.

Bellinger, R.A. (2013). The Géwël Tradition Project: Supporting a living tradition. *African Arts, 46*(1), 62–71. http://dx.doi.org/10.1162/AFAR_a_00045

Benckendorff, P., & Pearce, P. (2012). The psychology of events. In S.J. Page and J. Connell (Eds.), *The Routledge Handbook of Events*, 165–85. Milton Park, UK: Routledge.

Benfield, R.W. (Ed.). (2013). *Garden Tourism*. Wallingford, UK: CABI. http://dx.doi.org/10.1079/9781780641959.0000

Bennett, A., & Rogers, I. (2014). Street music, technology and the urban soundscape. *Continuum: Journal of Media and Cultural Studies, 28*(4), 454–64.

Benski, T., Langman, L., Perugorría, I., & Tejerina, B. (2013). From the streets and squares to social movement studies: What have we learned? *Current Sociology*, *61*(4), 541–61. http://dx.doi.org/10.1177/0011392113479753

Berg, M.L., & Sigona, N. (2013). Ethnography, diversity and urban space. *Identities: Global Studies in Culture and Power*, *20*(4), 347–60. http://dx.doi.org/10.1080/1070289X.2013.822382

Berridge, G. (2010). Event pitching: The role of design and creativity. *International Journal of Hospitality Management*, *29*(2), 208–15. http://dx.doi.org/10.1016/j.ijhm.2009.10.016

Best, J., & Horiuchi, G.T. (1985). The razor blade in the apple: The social construction of urban legends. *Social Problems*, *32*(5), 488–99. http://dx.doi.org/10.2307/800777

Betancur, J. (2011). Gentrification and community fabric in Chicago. *Urban Studies*, *48*(2), 383–406. http://dx.doi.org/10.1177/0042098009360680

Better Evaluation. (2005). *Community Fairs*. http://betterevaluation.org/evaluation-options/community_fairs

Biaett, V. (2012). A confessional tale: Auto-ethnography reflections on the investigation of attendee behavior at community festivals. *Tourism Today*, (Fall), 65–75.

Billig, S.H. (2012). Service-learning. In J. Hattie & E.M. Anderman (Eds.), *International Guide to Student Achievement*, 158–61. New York: Routledge.

Bishop, C. (2012). *Artificial Hells: Participatory Art and the Politics of Spectatorship*. London: Verso.

Blee, K., & McDowell, A. (2013). The duality of spectacle and secrecy: A case study of fraternalism in the 1920s US Ku Klux Klan. *Ethnic and Racial Studies*, *36*(2), 249–65. http://dx.doi.org/10.1080/01419870.2012.676197

Blešić, I., Pivac, T., Stamenković, I., & Besermenji, S. (2013). Motives of visits to ethno music festivals with regard to gender and age structure of visitors. *Event Management*, *17*(2), 145–54. http://dx.doi.org/10.3727/152599513X13668224082387

Blichfeldt, B.S., & Halkier, H. (2014). Mussels, tourism and community development: A case study of place branding through food festivals in rural North Jutland, Denmark. *European Planning Studies*, *22*(8), 1587–603.

Blickstein, S.G. (2010). Automobility and the politics of bicycling in New York City. *International Journal of Urban and Regional Research*, *34*(4), 886–905. http://dx.doi.org/10.1111/j.1468-2427.2010.00914.x

Blunt, A. (2007). Cultural geographies of migration: Mobility, transnationality and diaspora. *Progress in Human Geography*, *31*(5), 684–94. http://dx.doi.org/10.1177/0309132507078945

Bochenek, M. (2013). Festival tourism of folk group dancers from selected countries of the world. *Polish Journal of Sport and Tourism, 20*(2), 95–9. http://dx.doi.org/10.2478/pjst-2013-0009

Boersma, K. (2013). Liminal surveillance: An ethnographic control room study during a local event. *Surveillance & Society, 1*(1–2), 106–20.

Bogad, L.M. (2010). Carnivals against capital: Radical clowning and the global justice movement. *Social Identities: Journal for the Study of Race, Nature and Culture, 16*(4), 537–57.

Bonilla, M.H. (2012). The (re)construction of public space in today's Mexican City. *ArchNet-IJAR: International Journal of Architectural Research, 6*(2).

Boo, S., Carruthers, C.P., & Busser, J.A. (2014). The constraints experienced and negotiation strategies attempted by nonparticipants of a festival event. *Journal of Travel & Tourism Marketing, 31*(2), 269–85. http://dx.doi.org/10.1080/10548408.2014.873317

Boo, H.C., & Chan, W.L. (2014). Food safety at fairs and festivals. In B. Almanza & R. Ghiselli (Eds.), *Food Safety: Researching the Hazard in Hazardous Foods*, 341–67. Waretown, NJ: Apple Academic Press.

Botelho-Nevers, E., & Gautret, P. (2012). Outbreaks associated with large open air festivals, including music festivals, 1980–2012. *Bulletin européen sur les maladies transmissibles/European Communicable Disease Bulletin, 18*(11), 204–26.

Bottrell, D. (2009). Dealing with disadvantage: Resilience and the social capital of young people's networks. *Youth & Society, 40*(4), 476–501. http://dx.doi.org/10.1177/0044118X08327518

Bowdin, G., Allen, J., Harris, R., McDonnell, I., & O'Toole, W. (2012). *Events Management*. New York: Routledge.

Bowditch, R. (2013). Phoenix rising: The culture of fire at the Burning Man Festival. *Performance Research: A Journal of the Performing Arts, 18*(1), 113–22. http://dx.doi.org/10.1080/13528165.2013.789257

Bowen, H., & Daniels, M.J. (2005). Does the music matter? Motivations for attending a music festival. *Event Management, 9*(3), 155–64. http://dx.doi.org/10.3727/152599505774791149

Bowman, A.O.M., & Pagano, M.A. (2010). *Terra Incognita: Vacant Land and Urban Strategies*. Washington, DC: Georgetown University Press.

Bowmile, M. (2009). *From Queen Street to Queen's Park – The Word on the Street and festival culture*. Toronto: Robarts Centre for Canadian Studies, York University.

Bramadat, P.A. (2004). Mirror and mortar: Ethno-cultural festivals and urban life in Canada. In C. Andrew (Ed.), *Our Diverse Cities*, 87–91. Ottawa: Metropolis.

Brannstrom, C., & Brandao, P.R.B. (2012). Two hundred hectares of good business: Brazilian agriculture in a themed space. *Geographical Review, 102*(4), 465–85. http://dx.doi.org/10.1111/j.1931-0846.2012.00170.x

Braquet, D., & Westfall, M. (2011). Of fairs and festivals: Librarians teach thematic first-year seminars. *Southeastern Librarian, 59*(1), art3.

Bray, T.J. (2011, July). Orthopaedic vendor fairs. *Journal of Orthopaedic Trauma.*

Bressan, A., & Alonso, A.D. (2013). Cultural institutes as social anchors: Implications for tourism and hospitality planning and development. *Tourism Planning & Development, 10*(4), 433–50. http://dx.doi.org/10.1080/21568316.2013.779315

Brickell, C. (2000). Heroes and invaders: Gay and lesbian pride parades and the public/private distinction in New Zealand media accounts. *Gender, Place and Culture, 7*(2), 163–78. http://dx.doi.org/10.1080/713668868

Brown, A., Lyons, M., & Dankoco, I. (2010). Street traders and the emerging spaces for urban voice and citizenship in African cities. *Urban Studies, 47*(3), 666–83. http://dx.doi.org/10.1177/0042098009351187

Brown, M. (2014). Gender and sexuality II: There goes the gayborhood? *Progress in Human Geography, 38*(3), 457–65. http://dx.doi.org/10.1177/0309132513484215

Brown, S. (2014). Emerging professionalism in the event industry: A practitioner's perspective. *Event Management, 18*(1), 15–24. http://dx.doi.org/10.3727/152599514X13883555341760

Brown, S., & James, J. (2004). Event design and management: Ritual sacrifice? In I. Yeoman, M. Robertson, J. Ali-Knight, S. Drummond, & U. McMahan-Beattie (Eds.), *Festival and Events Management: An International Arts and Culture Perspective,* 53–64. Oxford: Butterworth-Heinemann.

Brown, S.J.W., & Trimboli, D. (2011). The real "worth" of festivals: Challenges for measuring socio-cultural impacts. *Asia Pacific Journal of Arts and Cultural Management, 8*(1), 616–29.

Browne, K. (2007). Lesbian geographies. *Social & Cultural Geography, 8*(1), 1–7. http://dx.doi.org/10.1080/14649360701251486

Bruce, K.M. (2013). LGBT Pride as a cultural protest tactic in a southern city. *Journal of Contemporary Ethnography, 42*(5), 608–35. http://dx.doi.org/10.1177/0891241612474933

Brueggemann, W.G. (2013). History and context for community practice in North America. In M. Weil, M. Reisch, & M.L. Ohmer (Eds.), *The Handbook of Community Practice,* 27–46. Los Angeles: Sage. http://dx.doi.org/10.4135/9781412976640.n2

Bruhn, M., Schoenmuller, V., Schafer, D., & Heinrich, D. (2012). Brand authenticity: Towards a deeper understanding of its conceptualization and measurement. *Advances in Consumer Research, 40*(4), 567–76.

Buch, T., Milne, S., & Dickson, G. (2011). Multiple stakeholder perspectives on cultural events: Auckland's Pasifika Festival. *Journal of Hospitality Marketing &*

Management, 20(3–4), 311–28. http://dx.doi.org/10.1080/19368623.2011.562416

Buckley, N., McPhee, J. & Jensen, E. (2011). *University Engagement in Festivals: Tips and Case Studies*. Cambridge, UK: National Co-ordinating Centre for Public Engagement. http://www.publicengagement.ac.uk/sites/default/files/publication/festivals_guides_and_top_tips_0.pdf

Bueltmann, T. (2012). "The image of Scotland which we cherish in our hearts": Burns anniversary celebrations in colonial Otago. *Immigrants & Minorities: Historical Studies in Ethnicity, Migration and Diaspora, 30*(1), 78–97.

Bueno, M.S., & Milanese, G. (2012). Hospitality and commensality at street fairs in the city of São Paulo: Fair Kantuta Cultural Boliviana. *Turismo y Desarrollo: Revista de Investigación en Turisme y Desarrollo Local. 5*(13). http://www.eumed.net/rev/turydes/13/feiras-rua-cidade-sao-paulo.pdf

Buggenhagen, B. (2014). A snapshot of happiness: Photo albums, respectability and economic uncertainty in Dakar. *Africa: Journal of the International African Institute, 84*(1), 78–100. http://dx.doi.org/10.1017/S0001972013000612

Burch, G.W. (2014). Child protagonism in transformational community development. *Transformation: An International Journal of Holistic Mission Studies, 31*(1), 36–46.

Burgess, A.H. (2011). *It's Not A Parade, It's A March!: Subjectivities, Spectatorship, and Contested Spaces of the Toronto Dyke March*. Doctoral dissertation, University of Toronto.

Burron, A., & Chapman, L.S. (2011). The use of health fairs in health promotion. *American Journal of Health Promotion, 25*(6), 1–12.

Butticci, A. (2013). Lagos Redemption: Dread, Crowd, and Charisma. Social Science Research Network. http://ssrn.com/abstract=2253758

Cabral, S., Krane, D., & Dantas, F. (2011). Carnaval in Salvador, Brazil: A case study of inter-organizational collaboration. Paper presented at JCPA/ICPA-Forum Comparative Research Symposium on Public Policy in Brazil and Latin America, Fundação Getulio Vargas, 18–19 Nov., São Paulo, Brazil.

Calestani, M. (2013). *An Anthropology Journey into Well-Being: Insights from Bolivia*. Dordrecht: Springer.

Calley Jones, C., & Mair, H. (2014). Magical activism: What happens between the worlds changes the worlds. *Annals of Leisure Research, 17*(3), 296–313.

Campbell, S.D. (2013). Sustainable development and social justice: Conflicting urgencies and the search for common ground in urban and regional planning. *Michigan Journal of Sustainability, 1*.

Cao, T.Q. (2011). *Celebratory Parade Violence: An Exploratory Study of the Role of Impression Management and Deindividuation*. Doctoral Dissertation, Northeastern University, Boston.

Caramellino, G., De Magistris, A., & Deambrosis, F. (2011). Reconceptualizing mega events and urban transformations in the twentieth century. *Planning Perspectives*, *26*(4), 617–20. http://dx.doi.org/10.1080/02665433.2011.599930

Cardoso, R. (2013). Building senses of" community": Social memory, popular movements and political participation. *Vibrant: Virtual Brazilian Anthropology*, *10*(1), 134–44.

Carlsen, J. (2004). The economics and evaluation of festivals and events. In I. Yeoman, M. Robertson, J. Ali-Knight, S. Drummond, & U. McMahan-Beattie (Eds.), *Festival and Events Management: An International Arts and Culture Perspective*, 246–59. Oxford: Butterworth-Heinemann. http://dx.doi.org/10.1016/B978-0-7506-5872-0.50021-1

Carlsen, J., Andersson, T.D., Ali-Knight, J., Jaeger, K., & Taylor, R. (2010). Festival management innovation and failure. *International Journal of Event and Festival Management*, *1*(2), 120–31. http://dx.doi.org/10.1108/17852951011056900

Carlsen, J., Getz, D., & Soutar, G. (2000). Event evaluation research. *Event Management*, *6*(4), 247–57. http://dx.doi.org/10.3727/152599500108751408

Carlsen, J., Robertson, M., & Ali-Knight, J. (2007). Social dimensions of community festivals: An application of factor analysis in the development of the social impact perception (SIP) scale. *Event Management*, *11*(1/2), 45–55.

Carmody, P.C. (2013). *Understanding Meaning and Existence: Toward the Development of a Measure of Existential Authenticity*. Doctoral dissertation, University of Tennessee.

Carney, J.K., Maltby, H.J., Mackin, K.A., & Maksym, M.E. (2011). Community-academic partnerships: How can communities benefit? *American Journal of Preventive Medicine*, *41*(4), S206–S213. http://dx.doi.org/10.1016/j.amepre.2011.05.020

Carson, A.J., Chappell, N.L., & Knight, C.J. (2007). Promoting health and innovative health promotion practice through a community arts centre. *Health Promotion Practice*, *8*(4), 366–374. http://dx.doi.org/10.1177/1524839906289342

Carter, R.D., & Zieren, J.W. (2012, 27 June). Festivals that Say CHA-CHING! Measuring the Economic Impact of Special Events. *Main Street Now*. Washington, DC: National Main Street Center.

Carter, S.B. (2013). *Embracing Isolation: Chinese American Geographic Redistribution*. http://mohandasmohandas.com/african5/Carter.pdf

Castéran, H., & Roederer, C. (2013). Does authenticity really affect behavior? The case of the Strasbourg Christmas Market. *Tourism Management*, *36*(2), 153–63. http://dx.doi.org/10.1016/j.tourman.2012.11.012

Castro, F.G., Barrera, M., Jr., Mena, L.A., & Aguirre, K.M. (2014). Culture and alcohol use: Historical and sociocultural themes from 75 years of alcohol research. *Journal of Studies on Alcohol and Drugs*, (S17): 36–49. http://dx.doi.org/10.15288/jsads.2014.s17.36

Castro, L.A., & Gonzalez, V.M. (2014). Transnational imagination and social practices: A transnational website in a migrant community. *Human-Computer Interaction*, 29(1), 22–52. http://dx.doi.org/10.1080/07370024.2013.823820

Caust, J., & Glow, H. (2011). Festivals, artists and entrepreneurialism: The role of the Adelaide Fringe Festival. *International Journal of Event Management Research*, 6(2), 1–14.

Chacko, H.E., & Schaffer, J.D. (1993). The evolution of a festival: Creole Christmas in New Orleans. *Tourism Management*, 14(6), 475–82. http://dx.doi.org/10.1016/0261-5177(93)90100-Y

Chafets, Z.E. (2013). *Devil's Night: And Other True Tales of Detroit*. New York: Random House.

Chan, F.H.-H. (2013). Intercultural climate and belonging in the globalizing multiethnic neighborhoods of Los Angeles. *Open Urban Studies Journal*, 6(1), 30–9.

Chandler, E. (2013). Mapping difference: Critical connections between crip and diaspora communities. *Critical Disability Discourse/Discours Critiques dans le Champ du Handicap*, 5, 39–66. http://cdd.journals.yorku.ca/index.php/cdd/article/view/37455/33996

Chang, S., & Mahadevan, R. (2014). Fad, fetish or fixture: Contingent valuation of performing and visual arts festivals in Singapore. *International Journal of Cultural Policy*, 20(3), 318–40. http://dx.doi.org/10.1080/10286632.2013.817396

Chang, W.H., Chang, K.S., Huang, C.S., Huang, M.Y., Chien, D.K., & Tsai, C.H. (2010). Mass gathering emergency medicine: A review of the Taiwan experience of long-distance swimming across Sun-Moon Lake. *International Journal of Gerontology*, 4(2), 53–68.

Chapple, K., & Jackson, S. (2010). Commentary: Arts, neighborhoods, and social practices – Towards an integrated epistemology of community arts. *Journal of Planning Education and Research*, 29(4), 478–90. http://dx.doi.org/10.1177/0739456X10363802

Charles B. Wang Health Center. (2012). *Our History*. http://www.cbwchc.org/history.asp

Chaskin, R.J. (2013). Theories of community. In M. Weil, M.S. Reisch, and M.L. Omner, *Handbook of community practice*, 105–21. Thousand Oaks, CA: Sage Publications.

Chhabra, D. (2010). Branding authenticity. *Tourism Analysis*, 15(6), 735–40. http://dx.doi.org/10.3727/108354210X12904412050134

Chiodelli, F., & Moroni, S. (2014). Typology of spaces and topology of toleration: City, pluralism, ownership. *Journal of Urban Affairs, 36*(2), 167–81.

Chirieleison, C., Montrone, A., & Scrucca, L. (2013). Measuring the impact of a profit-oriented event on tourism: The Eurochocolate Festival in Perugia, Italy. *Tourism Economics, 19*(6), 1411–28. http://dx.doi.org/10.5367/te.2013.0269

Choi, J.K., & Almanza, B. (2012). An assessment of food safety risk at fairs and festivals: A comparison of health inspection violations between fairs and festivals and restaurants. *Event Management, 16*(4), 295–303. http://dx.doi.org/10.3727/152599512X13539583374974

Choudhry, V., Agardh, A., Stafstrom, M., & Ostergren, P.-O. (2014). Patterns of alcohol consumption and risky sexual behavior: A cross-sectional study among Ugandan university students. *BMC Public Health, 14*(1), 128. http://dx.doi.org/10.1186/1471-2458-14-128

Choy, P.P. (2012). *San Francisco Chinatown: A Guide to Its History and Its Architecture.* San Francisco: City Lights Books.

Cicea, C., & Pirlogea, C. (2011). Green spaces and public health in urban areas. *Theoretical and Empirical Researches in Urban Management, 6*(1), 83–92.

Clarke, A., & Jepson, A. (2011). Power and hegemony within a community festival. *International Journal of Event and Festival Management, 2*(1), 7–19. http://dx.doi.org/10.1108/17582951111116588

Clarke, V., Burgoyne, C., & Burns, M. (2013). Unscripted and improvised: Public and private celebrations of same-sex relationships. *Journal of GLBT Family Studies, 9*(4), 393–418. http://dx.doi.org/10.1080/1550428X.2013.808494

Clifton, N., O'Sullivan, D., & Pickernell, D. (2012). Capacity building and the contribution of public festivals: Evaluating "Cardiff 2005." *Event Management, 16*(1), 77–91. http://dx.doi.org/10.3727/152599512X13264729827712

Clopton, A.W., & Finch, B.L. (2011). Re-conceptualizing social anchors in community development: Utilizing social anchor theory to create social capital's third dimension. *Community Development, 42*(1), 70–83. http://dx.doi.org/10.1080/15575330.2010.505293

Coffman, E. (2009). Documentary and collaboration: Placing the camera in the community. *Journal of Film and Video, 61*(1), 62–78. http://dx.doi.org/10.1353/jfv.0.0017

Cohen, E., & Cohen, S.A. (2012). Authentication: Hot and cool. *Annals of Tourism Research, 39*(3), 1295–314. http://dx.doi.org/10.1016/j.annals.2012.03.004

Colding, J., & Barthel, S. (2013). The potential of "urban green commons" in the resilience building of cities. *Ecological Economics, 86*(2), 155–66.

Cole, S., & Morgan, N. (Eds.). (2010). *Tourism and Inequality: Problems and Prospects*. Wallingford, UK: CABI. http://dx.doi.org/10.1079/9781845936624.0000

Collin-Lachaud, I., & Kjeldgaard, D. (2013). Loyalty in a cultural perspective: Insights from French music festivals. *Research in Consumer Behavior, 15*(4), 285–95. http://dx.doi.org/10.1108/S0885-2111(2013)0000015019

Collins, D., Parsons, M., & Zinyemba, C. (2014). Air quality at outdoor community events: Findings from fine particulate (PM 2.5) sampling at festivals in Edmonton, Alberta. *International Journal of Environmental Health Research, 24*(3), 215–25. http://dx.doi.org/10.1080/09603123.2013.807328

Collins, R. (2013). Public festivals: Ritual successes, failures and mediocrities. *Polis, 1*(Apr.), 13–28.

Collura, J.J., & Christens, B.D. (2015). Perspectives on systems change among local change agents: A comparative study. *Journal of Community & Applied Social Psychology, 25*(1), 19–33.

Congeong, T. (2014). The study of festival tourism development in Shanghai. *International Journal of Business and Social Science, 5*(4), 52–8.

Conklin-Ginop, E., Braverman, M.T., Caruso, R., & Bone, D. (2011). Bringing Carnaval drum and dance traditions into 4-H programming for Latino youth. *Journal of Extension, 49*(4), art4IAW1.

Connolly, T.H. (2010). Business ritual studies: Corporate ceremony and sacred space. *International Journal of Business Anthropology, 1*(2), 32–47.

Cornwell, B., & Warburton, E. (2014). Work schedules and community ties. *Work and Occupations, 41*(2), 139–74. http://dx.doi.org/10.1177/0730888413498399

Correll, T.C. (2014). Productos Latinos: Latino business murals, symbolism, and the social enactment of identity in Greater Los Angeles. *Journal of American Folklore, 127*(505), 285–320. http://dx.doi.org/10.5406/jamerfolk.127.505.0285

Coser, L.R., Tozar, K., Borek, N.V., Tzemis, D., Taylor, D., & Saewyc, E. (2014). Finding a voice: Participatory research with street-involved youth in the Youth Injection Prevention Project. *Health Promotion Practice, 15*(5), 732–8.

Costanzo, J.M. (2012). *Practicing Local Culture as a Vehicle of Integration? Creative Collaborations and Brussels' Zinneke Parade*. Doctoral dissertation, University of Maryland.

Couser, W.G., Shah, S., Kopple, J., Beerkens, P., Wilson, A., Feehally, J., ..., & Riella, M. (2007). A call to action on World Kidney Day, 8 March 2007. *Kidney International, 71*(5), 369–70. http://dx.doi.org/10.1038/sj.ki.5002144

Crespi-Vallbona, M., & Richards, G. (2007). The meaning of cultural festivals. *International Journal of Cultural Policy, 13*(1), 103–22. http://dx.doi.org/10.1080/10286630701201830

Croes, R., Lee, S.H., & Olson, E.D. (2013). Authenticity in tourism in small island destinations: A local perspective. *Journal of Tourism and Cultural Change, 11*(1–2), 1–20. http://dx.doi.org/10.1080/14766825.2012.759584

Crompton, J.L., & McKay, S.L. (1997). Motives of visitors attending festival events. *Annals of Tourism Research, 24*(2), 425–39. http://dx.doi.org/10.1016/S0160-7383(97)80010-2

Crooks, R.N. (2013). The rainbow flag and the green carnation: Grindr in the gay village. *First Monday: Peer-Reviewed Journal on the Internet, 18*(11).

Crowther, P. (2010). Strategic application of events. *International Journal of Hospitality Management, 29*(2), 227–35. http://dx.doi.org/10.1016/j.ijhm.2009.10.014

Croy, W.G. (2010). Planning for film tourism: Active destination image management. *Tourism and Hospitality Planning & Development, 7*(1), 21–30. http://dx.doi.org/10.1080/14790530903522598

Cudny, W. (2014). Festivals as a subject for geographical research. *Geografisk Tidsskrift-Danish Journal of Geography, 114*(2), 132–42. http://dx.doi.org/10.1080/00167223.2014.895673

Cumming, D. (2012). Putting your community on stage: Creating a play out of local historical letters to the editor. *American Journalism, 29*(2), 141–6.

Cunningham, F. (2011). The virtues of urban citizenship. *City, Culture, & Society, 2*(1), 35–44.

Cunningham, N., & Gregory, I. (2014). Hard to miss, easy to blame? Peacelines, interfaces and political deaths in Belfast during the Troubles. *Political Geography, 40*(1), 64–78. http://dx.doi.org/10.1016/j.polgeo.2014.02.004

Cupertino, A.P., Cox, L.S., Garrett, S., Suarez, N., Sandt, H., Mendoza, I., & Ellerbeck, E.F. (2011). Tobacco use and interest in smoking cessation among Latinos attending community health fairs. *Journal of Immigrant Health, 13*(4), 719–24. http://dx.doi.org/10.1007/s10903-010-9404-y

Curtis, R.A. (2010). Australia's capital of jazz? The (re)creation of place, music and community at the Wangaratta Jazz Festival. *Australian Geographer, 41*(1), 101–16. http://dx.doi.org/10.1080/00049180903535618

Cutchin, M.P., Eschbach, K., Mair, C.A., Ju, H., & Goodwin, J.S. (2011). The socio-spatial neighborhood estimation method: An approach to operationalizing the neighborhood concept. *Health & Place, 17*(5), 1113–21. http://dx.doi.org/10.1016/j.healthplace.2011.05.011

Damm, S. (2011). *Event Management: How to Apply Best Practices to Small Scale Events.* Hamburg: Diplomica Verlag.

Daniels, J. (2012). Digital video: Engaging students in critical media literacy and community activism. *Explorations in Media Ecology, 10*(1–2).

Daniels, R. (1988). *Asian America: Chinese and Japanese in the United States since 1850.* Seattle: University of Washington Press.

Daniels, R. (2004). *Guarding the Golden Door: American Immigration Policy and Immigrants since 1882.* New York: Hill and Wang.

Daniels, T.P. (2010). Urban space, belonging, and inequality in multi-ethnic housing estates of Melaka, Malaysia. *Identities: Global Studies in Culture and Power, 17*(2–3), 176–203. http://dx.doi.org/10.1080/10702891003734953

Dao, L., Morrison, L., & Zhang, C. (2012). Bonfires as a potential source of metal pollutants in urban soils, Galway, Ireland. *Applied Geochemistry, 27*(4), 930–5. http://dx.doi.org/10.1016/j.apgeochem.2012.01.010

Darian-Smith, K. (2011). Histories of Agricultural Shows and Rural Festivals in Australia. In C. Gibson & J. Connell (Eds.), *Festival Places: Revitalising Rural Australia*, 25–43. Bristol: Channel View Publications.

Davies, L., Coleman, R., & Ramchandani, G. (2013). Evaluating event economic impact: Rigour versus reality? *International Journal of Event and Festival Management, 4*(1), 31–42. http://dx.doi.org/10.1108/17582951311307494

Davilla, A. (2012). *Culture Works: Space, Value, and Mobility across the Neoliberal Americas.* New York: New York University Press.

Davis, G.F. (2009). *Anzac Day Meanings and Memories: New Zealand, Australian and Turkish Perspectives on a Day of Commemoration in the Twentieth Century.* Doctoral dissertation, University of Otago.

Davis, M. (2001). *Magical Urbanism: Latinos Reinvent the U.S. Big City.* New York: Verso.

Dawson, E., & Jensen, E. (2011). Towards a contextual turn in visitor studies: Evaluating visitor segmentation and identity-related motivation. *Visitor Studies, 14*(2), 127–40. http://dx.doi.org/10.1080/10645578.2011.608001

del Barrio, M.J., Devesa, M., & Herrero, L.C. (2012). Evaluating intangible cultural heritage: The case of cultural festivals. *City, Culture and Society, 3*(4), 235–44.

De Bres, K., & Davis, J. (2001). Celebrating group and place identity: A case study of a new regional festival. *Tourism Geographies: An International Journal of Tourism Space, Place and Environment, 13*(3), 326–37.

Deery, M., & Jago, L. (2010). Social impacts of events and the role of anti-social behavior. *International Journal of Event and Festival Management, 1*(1), 8–28. http://dx.doi.org/10.1108/17852951011029289

Delacour, H., & Leca, B. (2011). A Salon's life: Field-configuring event, power and contestation in a creative field. In B. Moeran & J.S. Pedersen (Eds.), *Negotiating Values in the Creative Industries: Fairs, Festivals and Competitive Events*, 36–58. New York: Cambridge University Press.

Delgado, M. (1999a). *Social Work Practice in Nontraditional Urban Settings.* New York: Oxford University Press.

Delgado, M. (1999b). *Community Social Work Practice in an Urban Context: The Potential of a Capacity-Enhancement Perspective.* New York: Oxford University Press.

Delgado, M. (2003). *Death at an Early Age and the Urban Scene: The Case for Memorial Murals and Community Healing.* Westport, CT: Praeger.

Delgado, M. (2007). *Social Work Practice with Latinos: A Cultural Assets Paradigm.* New York: Oxford University Press.

Delgado, M. (2012). *Latino Small Businesses and the American Dream: Community Social Work Practice and Economic and Social Development.* New York: Columbia University Press.

Delgado, M. (2015). *Urban Youth Photovoice: Visual Ethnography in Action.* New York: Oxford University Press.

Delgado, M. (2016). *Community Practice and Urban Youth: Social Justice Service-Learning and Civic Engagement.* New York: Routledge.

Delgado, M., & Humm-Delgado, D. (2013). *Asset Assessments and Community Social Work Practice.* New York: Oxford University Press. http://dx.doi.org/10.1093/acprof:oso/9780199735846.001.0001

Delgado, M., Jones, K., & Rohani, M. (2005). *Social Work Practice with Immigrant and Refugee Youth in the United States.* Boston: Allyn & Bacon.

Delgado, M., & Staples, L. (2008). *Youth-Led Community Organizing: Theory and Action.* New York: Oxford University Press.

Dempsey, N., Bramley, G., Power, S., & Brown, C. (2011). The social dimension of sustainable development: Defining urban social sustainability. *Sustainable Development, 19*(5), 289–300. http://dx.doi.org/10.1002/sd.417

Deng, J., & Pierskalla, C. (2011). Impact of past experience on perceived value, overall satisfaction, and destination loyalty: A comparison between visitor and resistant attendees of a festival. *Event Management, 15*(2), 163–77. http://dx.doi.org/10.3727/152599511X13082349958235

DePalma, R. (2009). Leaving Alinsu: Towards a transformative community practice. *Mind, Culture, and Activity, 16*(4), 353–70. http://dx.doi.org/10.1080/10749030902818394

Derrett, R. (2003). Making sense of how festivals demonstrate a community's sense of place. *Event Management, 8*(1), 49–58. http://dx.doi.org/10.3727/152599503108751694

Derrett, R. (2004). Festivals, events and the destination. In I. Yeoman, M. Robertson, J. Ali-Knight, S. Drummond & U. McMahan-Beattle (Eds.), *Festival and Events Management: An International Arts and Culture Perspective*, 32–49. Oxford: Butterworth-Heinemann.

Devesa, M., Baez, A., Figueroa, V., & Herrero, L.C. (2012). Repercusiones económicas y sociales de los festivales culturales: el caso del Festival Internacional de Cine de Valdivia. [Santiago]. *EURE: Revista Latinoamericana de Estudios Urbano Regionales, 38*(115), 95–115. http://dx.doi.org/10.4067/S0250-71612012000300005

Devine, L., Moss, S., & Walmsley, B. (2014). Event planning and management. In S. Moss & B. Walmsley (Eds.), *Entertainment Management: Towards Best Practice*, 99. Wallingford, UK: CABI.

Dias, D.R. (2005). *Barrio Urbanism: Chicanos, Planning, and American Cities*. New York: Routledge.

Dickson, C., & Arcodia, C. (2010). Promoting sustainable event practice: The role of professional associations. *International Journal of Hospitality Management, 29*(2), 236–44. http://dx.doi.org/10.1016/j.ijhm.2009.10.013

Diehl, D., & Donnelly, M. (2011). *Medieval Celebrations: Your Guide to Planning and Hosting Spectacular Feasts, Parties, Weddings, and Renaissance Fairs*. Mechanicsburg, PA: Stackpole Books.

Dillon, D.L., & Sternas, K. (1997). Designing a successful health fair to promote individual, family, and community health. *Journal of Community Health Nursing, 14*(1), 1–14. http://dx.doi.org/10.1207/s15327655jchn1401_1

Dimitrova, M., & Yoveva, I. (2014). Food festivals as purpose-created tourism attractions and their impact on destination branding: Lessons learned from the development of a regional food festival in Bulgaria. In A. Cavicchi & C. Santini (Eds.), *Food and Wine Events in Europe: A Stakeholder Approach*, 172–81. New York: Routledge.

Dinaburgskava, K., & Ekner, P. (2010). *Social impacts of the Way Out Festival on the residents of the City of Goteborg*. Master of science thesis, University of Gothenburg, Sweden.

Disegna, M., Brida, J.G., & Osti, L. (2011). Authenticity Perception of Cultural Events: A Host-Tourist Analysis. *Tourism, Culture & Communication, 12*(2).

Doan, P.L., & Higgins, H. (2011). The demise of queer space? Resurgent gentrification and the assimilation of LGBT neighborhoods. *Journal of Planning Education and Research, 31*(1), 6–25. http://dx.doi.org/10.1177/0739456X10391266

Dodd, T., Yuan, J., Adams, C., & Kolyesnikova, N. (2006). Motivations of young people for visiting wine festivals. *Event Management, 10*(1), 23–33. http://dx.doi.org/10.3727/152599506779364660

Dodds, R., & Graci, S. (2012). Greening of the Pride Toronto Festival: Lessons learned. *Tourism Culture & Communication, 12*(1), 29–38. http://dx.doi.org/10.3727/109830412X13542041184739

Donaldson, L.P., & Daughtery, L. (2011). Introducing asset-based models of social justice into service learning: A social work approach. *Journal of Community Practice*, *19*(1), 80–99. http://dx.doi.org/10.1080/10705422.2011.550262

Donish, C. (2013). *Geographies of Art and Urban Change: Contesting Gentrification through Aesthetic Encounters in San Francisco's Mission District*. Doctoral dissertation, University of Oregon.

Douglas, P. (2007, Sept.-Oct.). Diversity and the gay and lesbian community: More than chasing the pink dollar. *Ivey Business Journal Online*.

Dressler, A.E., Scheibel, R.P., Wardyn, S., Harper, A.L., Hanson, B.M., Kroeger, J.S., ..., & Smith, T.C. (2012). Prevalence, antibiotic resistance and molecular characterisation of *Staphylococcus aureus* in pigs at agricultural fairs in the USA. *Veterinary Record*, *170*(19), 495.

Duffy, M., Waitt, G., Gorman-Murray, A., & Gibson, C. (2011). Bodily rhythms: Corporeal capacities to engage with festival spaces. *Emotion, Space and Society*, *4*(1), 17–24. http://dx.doi.org/10.1016/j.emospa.2010.03.004

Duguay, J.P. (2013). *Counteracting Stereotypes and Forgiving Bad Deeds: How Are Calgarian Journalists Framing the White Pride Parade?* Doctoral dissertation, Saint Paul University.

Dunlap, R., Harmon, J., & Kyle, G. (2013). Growing in place: The interplay of urban agriculture and place sentiment. *Leisure/Loisir*, *37*(4), 397–414.

Dunn-Young, F.A. (2012). *A Community Capitals Approach to Assessing Immigrant Access to Community Resources within a Meatpacking Community*. Dissertation, Iowa State University, Paper 12913.

Duran, E., Hamarat, B., & Özkul, E. (2014). A sustainable festival management model: The case of International Troia festival. *International Journal of Culture, Tourism and Hospitality Research*, *8*(2), 173–93.

Durkheim, E. (1915). *The Elementary Forms of the Religious Life*. London: Allen & Unwin.

Dzombic, J. (2014). Rightwing extremism in Serbia. *Race & Class*, *55*(4), 106–10. http://dx.doi.org/10.1177/0306396813519943

Eberle, S.G. (2014). The elements of play: Toward a philosophy and a definition of play. *American Journal of Play*, *6*(2), 214–33.

Ebert, K., & Okamoto, D.G. (2013). Social citizenship, integration and collective action: Immigrant civic engagement in the United States. *Social Forces*, *91*(4), 1267–92.

Ecoma, C.S., & Ecoma, L.E. (2013). Edele festival of the Itigid people of Cross River state and its impact. *International Journal of Asian Social Science*, *3*(8), 1781–92.

Edwards, G. (2011). Concepts of community: A framework for contextualizing distributed leadership. *International Journal of Management Reviews, 13*(3), 301–312. http://dx.doi.org/10.1111/j.1468-2370.2011.00309.x

Edwards, J., & Knottnerus, J.D. (2010). The Orange Order: Parades, other rituals, and their outcomes. *Sociological Focus, 43*(1), 1–23. http://dx.doi.org/10.1080/00380237.2010.10571366

Ehrkamp, P. (2008). Risking publicity: Masculinities and the racialization of public neighborhood space. *Social & Cultural Geography, 9*(2), 117–33. http://dx.doi.org/10.1080/14649360701856060

Eizenberg, E. (2012). Actually existing commons: Three moments of space of community gardens in New York City. *Antipode: A Radical Journal of Geography, 44*(3), 764–82.

Elgenius, G. (2011). *Symbols of Nations and Naturalism: Celebrating Nationhood.* Basingstoke: Palgrave Macmillan.

Elia, N. (2012). Gay rights with a side of Apartheid. *Settler Colonial Studies, 2*(2), 49–68. http://dx.doi.org/10.1080/2201473X.2012.10648841

Elias-Varotsis, S. (2006). Festivals and events – (Re) interpreting cultural identity. *Tourism Review, 61*(2), 24–29. http://dx.doi.org/10.1108/eb058472

Elsayad, N.A. (2014). Children's playgrounds & recreation areas. *Journal of Humanities and Social Science, 19*(3), 45–53.

Embassy of the United States Serbia. (2013). *Pride Parade Events in Downtown Belgrade.* http://serbia.usembassy.gov/messages-and-notices-for-us-citizens/pride-events-2013.html

English, J.F. (2011). Festivals and the geography of culture: African cinema in the "world space" of its public. In G. Delanty, L. Giorgi, & M. Sassatelli (Eds.), *Festivals and the Cultural Public Sphere,* 63–78. London: Routledge.

Ensor, J., Robertson, M., & Ali-Knight, J. (2011). Eliciting the dynamics of leading a sustainable event: Key informant responses. *Event Management, 15*(4), 315–27. http://dx.doi.org/10.3727/152599511X13175676722483

Entwistle, J., & Rocamora, A. (2011). Between art and commerce: London Fashion Week as trade fair and fashion spectacle. In B. Moeran & J.S. Pedersen (Eds.), *Negotiating Values in the Creative Industries: Fairs, Festivals and Competitive Events,* 249–69. New York: Cambridge University Press. http://dx.doi.org/10.1017/CBO9780511790393.011

Epstein, W.M. (2013). *Empowerment as Ceremony.* New Brunswick, NJ: Transaction Publishers.

Erasmus, L.J.J. (2012). *Key Success Factors in Managing the Visitors' Experience at the Klein Karoo National Arts Festival.* Doctoral dissertation, North-West University, Potchefstroom Campus, South Africa.

Erol, A. (2012). Identity, migration and transnationalism: Expressive cultural practices of the Toronto Alevi community. *Journal of Ethnic and Migration Studies*, *38*(5), 833–49. http://dx.doi.org/10.1080/1369183X.2012.668025

Escoffery, C., Rodgers, K., Kegler, M.C., Haardörfer, R., Howard, D., Roland, K.B., ..., & Rodriguez, J. (2014). Key informant interviews with coordinators of special events conducted to increase cancer screening in the United States. *Health Education Research*, *29*(5), 730–9.

Evans, G. (2012). Hold back the night: *Nuit Blanche* and all-night events in capital cities. *Current Issues in Tourism*, *15*(1–2), 35–49. http://dx.doi.org/10.1080/13683500.2011.634893

Evans, J.J., & York, K.B. (2013). How the Mall means: An analysis of the National Mall as a cohesively built environment. *Perspectives on Political Science*, *42*(3), 117–30. http://dx.doi.org/10.1080/10457097.2013.793515

Eversole, R. (2012). Remaking participation: Challenges for community development practice. *Community Development Journal: An International Forum*, *47*(1), 29–41. http://dx.doi.org/10.1093/cdj/bsq033

Everts, J., Lahr-Kurten, M., & Watson, M. (2011). Practice matters! Geographical inquiry and theories of practice. *Erdkunde: Archive for Scientific Geography*, *65*(4), 323–34.

Ezeonwu, M., & Berkowitz, B. (2014). A collaborative communitywide health fair: The process and impacts on the community. *Journal of Community Health Nursing*, *31*(2), 118–29. http://dx.doi.org/10.1080/07370016.2014.901092

Fainstein, S. (2010). *The Just City*. Ithaca, NY: Cornell University Press.

Fairer-Wessels, F.A., & Malherbe, N. (2012). Sustainable urban event practice: The role of corporate sponsors within Gauteng. *Urban Forum*, *23*(1), 93–106.

Farina, A. (2014). *Soundscape Ecology: Principles, Patterns, Methods and Applications*. Dordrecht: Springer.

Feng, L. (2013). *Efficient People Movement through Optimal Facility Configuration and Operation*. Dissertation, University of Maryland. http://drum.lib.umd.edu/handle/1903/14829

Ferdinand, N., & Williams, N.L. (2013). International festivals as experience production systems. *Tourism Management*, *34*(1), 202–10. http://dx.doi.org/10.1016/j.tourman.2012.05.001

Ferman, D. (2013). A parade or a riot: A discourse analysis of two ethnic newspapers on the 2011 Marching Season in Northern Ireland. *Journal of Media and Religion*, *12*(2), 55–70. http://dx.doi.org/10.1080/15348423.2013.811367

Fernandes, G. (2010). Beyond the "politics of toil": Collective mobilization and individual activism in Toronto's Portuguese community, 1950s–1990s. *Urban History Review/Revue d'histoire urbaine*, *39*(1), 59–72.

Ferreira, M.E.P. de C. (2012). Reconceptualising public places of (in) equality: Sensing and creating layers of visibility. In *NOVA Graduate Conference in Social Sciences and Humanities*, 26–7 Nov., Lisbon. Repositorio de Nova.

Ferry, N.C. (2012). Rethinking the mainstream gay and lesbian movement beyond the classroom: Exclusionary results from inclusion-based assimilation politics. *JCT Online: Journal of Curriculum Theorizing, 28*(2).

Fields, K., & Stansbie, P. (2003). Festival and event catering operations. In I. Yeoman, M. Robertson, J. Ali-Knight, S. Drummond, & U. McMahon-Beattie (Eds.), *Festival and Event Management: An International Arts and Culture Perspective*, 171–82. Oxford: Butterworth-Heinemann.

Fillieule, O. (2012). The independent psychological effects of participation in demonstrations. *Mobilization: An International Quarterly, 17*(3), 235–48.

Fine, G.A. (2014). The Hinge: Civil society, group culture, and the interaction order. *Social Psychology Quarterly, 77*(1), 5–26. http://dx.doi.org/10.1177/0190272514522769

Finkel, R. (2009). A picture of the contemporary combined arts festival landscape. *Cultural Trends, 18*(1), 3–21. http://dx.doi.org/10.1080/09548960802651195

Finkel, R. (2010). Re-imaging arts festivals through a corporate lens: A case study of business sponsorship at the Henley Festival. *Managing Leisure, 15*(4), 237–50. http://dx.doi.org/10.1080/13606719.2010.508664

Fitzpatrick, A. (2013). *Together Old and Young (TOY)*. The TOY [Together Old and Young] Project Consortium: Intergenerational Learning Involving Young Children and Older People. Report. Leiden: The TOY Project.

Flachs, A. (2010). Food for thought: The social impact of community gardening in the greater Cleveland area. *EGJ: Electronic Green Journal, 1*(30).

Flecha, A.C., Lott, W., Lee, T.J., Moital, M., & Edwards, J. (2010). Sustainability of events in urban historic centers: The case of Ouro Preto, Brazil. *Tourism and Hospitality Planning & Development, 7*(2), 131–43. http://dx.doi.org/10.1080/14790531003798302

Flint, R.W. (2010). Seeking resilience in the development of sustainable communities. *Research in Human Ecology, 17*(1), 44–57.

Foley, M., McGillivray, D., & McPherson, G. (2012). *Event Policy: From Theory to Strategy*. New York: Routledge.

Foner, N. (Ed.). (2001). *New Immigrants in New York*. New York: Columbia University Press.

Fong, E., Luk, C., & Ooka, E. (2005). Spatial distribution of suburban ethnic businesses. *Social Science Research, 34*(1), 215–35. http://dx.doi.org/10.1016/j.ssresearch.2003.10.007

Forsyth, M. (2012). Lifting the lid on "The Community": Who has the right to control access to traditional knowledge and expressions of culture? *International Journal of Cultural Property*, *19*(1), 1–31. http://dx.doi.org/10.1017/S0940739112000021

Fowler, S., Venuto, M., Ramirez, V., Martinez, E., Pollard, J., Hixon, J., ..., & Lesser, J. (2013). The power of connection and support in community engagement: Las Mujeres Nobles de Harlandale. *Issues in Mental Health Nursing*, *34*(4), 294–7. http://dx.doi.org/10.3109/01612840.2012.719584

Fox, D., Gouthro, M.B., Morakabati, Y., & Brackstone, J. (2014). *Doing Events Research: From Theory to Practice*. New York: Routledge.

Foxall, A. (2014). Performing ethnic relations in Russia's North Caucasus: Regional spectacles in Stavropol' krai. *Central Asian Survey*, *33*(1), 47–61.

Fradi, H., & Dugelay, J.L. (2013). Crowd density map estimation based on feature tracks. Paper presented at the 15th International Workshop on Multimedia Signal Processing, 30 Sept. to 2 Oct. 2013, Pula, Italy.

Francesco, M.G. (2014). Peruvians in Paterson: The growth and establishment of a Peruvian American community within the multiethnic immigrant history of Paterson, New Jersey. *Journal of Urban History*, *40*(3), 497–513.

Fredline, L., Deery, M., & Jago, L. (2013). A longitudinal study of the impacts of an annual event on local residents. *Tourism Planning & Development*, *10*(4), 416–32. http://dx.doi.org/10.1080/21568316.2013.779314

Freeman, R.E., & Auster, E.R. (2011). Values, authenticity, and responsible leadership. *Journal of Business Ethics*, *98*(S1), 15–23. http://dx.doi.org/10.1007/s10551-011-1022-7

Fricke, C. (2013). Protocol, politics and popular culture: The independence jubilee in Gabon. *Nations and Nationalism*, *19*(2), 238–56. http://dx.doi.org/10.1111/nana.12018

Friedman, S. (2003). *Meeting & Event Planning for Dummies*. New York: For Dummies.

Frost, W., & Laing, J. (2013). Communicating persuasive messages through slow food festivals. *Journal of Vacation Marketing*, *19*(1), 67–74. http://dx.doi.org/10.1177/1356766712461403

Frostig, K. (2011). Arts activism: Praxis in social justice, critical discourse, and radical modes of engagement. *Art Therapy: Journal of the American Art Therapy Association*, *28*(2), 50–6. http://dx.doi.org/10.1080/07421656.2011.578028

Fujimoto, Y., Rentschler, R., Le, H., Edwards, D., & Hartel, C.E.J. (2014). Lessons learned from community organizations: Inclusion of people with disabilities and others. *British Journal of Management*, *25*(3), 518–37.

Gabbert, L. (2011). *Winter Carnival in a Western Town: Identity, Change, and the Good of the Community*. Logan, UT: Utah State University Press.

Gabryś-Barker, D. (2014). What's in a name? Naming habits in Polish and Portuguese food culture. In A. Łyda & K. Szcześniak (Eds.), *Awareness in Action: The Role of Consciousness in Language Acquisition*, 209–23. Heidelberg: Springer. http://dx.doi.org/10.1007/978-3-319-00461-7_14

Gallagher, A., & Pike, K. (2011). Sustainable management for maritime events and festivals. *Journal of Coastal Research, 61*(2), 158–65. http://dx.doi.org/10.2112/SI61-001.10

Gandy, S.K., Pierce, J., & Smith, A.B. (2009). Collaboration with community partners: Engaging teacher candidates. *Social Studies, 100*(1), 41–5. http://dx.doi.org/10.3200/TSSS.100.1.41-45

Ganeva-Raycheva, V. (2013). Migration and memory: Policies and practices in the construction of identity and heritage. *Our Europe: Ethnography, "Ethnology," Anthropology of Culture, 2*(1), 7–22.

Gapas, D.F. (2013). *Evaluating Social Sustainability in Plans for Inter-Cultural Cities*. Master's thesis, Virginia Commonwealth University. VCU Theses and Dissertations. Paper 3018. http://scholarscompass.vcu.edu/etd/3018

García, B. (2004). Urban regeneration, arts programming and major events. *International Journal of Cultural Policy, 10*(1), 103–18. http://dx.doi.org/10.1080/1028663042000212355

García, L., & Cantú, M. (2007). Processing Latinidad: Mapping Latino urban landscapes through Chicago ethnic festivals. *Latino Studies, 5*(3), 317–39. http://dx.doi.org/10.1057/palgrave.lst.8600259

Gardi, A. (2014). *Visitor Satisfaction at a Local Festival: An Importance-Performance Analysis of Oktoberfest*. Master's thesis, University of Waterloo.

Gaulin, C., Lê, M.L., & Kosatsky, T. (2010). *Church/Community Suppers: What Is the Evidence for Risk of Food-Borne Illness?* National Collaborating Centre for Environmental Health. http://www.ncceh.ca/sites/default/files/Church_Community_Suppers_Mar_2010_2.pdf

Gębarowski, M. (2012). Educational fairs as a form of promotion of higher education institutions. In *Marketing of Scientific and Research Organizations, Scientific Publications of the Institute of Aviation*. Warsaw, 93–103. Social Science Research Network. http://papers.ssrn.com/sol3/papers.cfm?abstract_id=2411785

Geiger, M.P.A., James, F., Sabino, M.P.H., Gertner, M.D., Eric, J., Patton, M.D., ..., & Llewellyn, J. (2013). Bienvenidos: The initial phase of organizational transformation to enhance cross-cultural health care delivery in a large health network. *International Journal of Organizational Diversity, 12*(4), 25–36.

Geller, J.D., Zuckerman, N., & Seidel, A. (2014). Service-learning as a catalyst for community development: How do community partners benefit from service-learning? *Education and Urban Society.* http://eus.sagepub.com/content/early/2014/01/05/0013124513514773

Gerodimos, R. (2010). *New Media, New Citizens: The Terms and Conditions of Online Youth Civic Engagement.* Doctoral dissertation, Bournemouth University.

Getz, D. (2002). Why festivals fail. *Event Management, 7*(4), 209–19. http://dx.doi.org/10.3727/152599502108751604

Getz, D. (2009). Policy for sustainable and responsible festival events: Institutionalization of a new paradigm. *Journal of Policy Research in Tourism, Leisure and Events, 1*(1), 61–78. http://dx.doi.org/10.1080/19407960802703524

Getz, D. (2010). The nature and scope of festival studies. *International Journal of Event Management Research, 5*(1), 1–47.

Getz, D. (2012). *Event Studies: Theory, Research and Policy for Planned Events.* London: Routledge.

Getz, D. (2015). The forms and functions of planned events: Past and future. In I. Yeoman, M. Robertson, U. McMahon-Beattie, E. Backer, & K.A. Smith (Eds.), *The Future of Events and Festivals,* 20–35. New York: Routledge.

Getz, D., & Andersson, T. (2010). Festival stakeholders: Exploring relationships and dependency through a four-country comparison. *Journal of Hospitality & Tourism Research, 34*(4), 531–56. http://dx.doi.org/10.1177/1096348010370862

Getz, D., Andersson, T., & Carlsen, J. (2010). Festival management studies: Developing a framework and priorities for comparative and cross-cultural research. *International Journal of Event and Festival Management, 1*(1), 29–59. http://dx.doi.org/10.1108/17852951011029298

Getz, D., & Frisby, W. (1988). Evaluating management effectiveness in community-run festivals. *Journal of Travel Research, 27*(1), 22–7. http://dx.doi.org/10.1177/004728758802700105

Getz, D., & Robinson, R. N. (2014). Foodies and food events. *Scandinavian Journal of Hospitality and Tourism, 14*(30), 315–30.

Ghasemzadeh, B. (2013). Cultural spaces and urban identity. *International Journal of Architecture and Urban Development, 3*(1). http://www.sid.ir/en/VEWSSID/J_pdf/1036020130110.pdf

Ghaziani, A., & Baldassarri, D. (2011). Cultural anchors and the organization of differences: A multi-method analysis of LGBT marches on Washington. *American Sociological Review, 76*(2), 179–206. http://dx.doi.org/10.1177/0003122411401252

Giannoulakis, C., Wang, C.H., & Gray, D. (2007). Measuring volunteer motivation in mega-sporting events. *Event Management, 11*(4), 191–200. http://dx.doi.org/10.3727/152599508785899884

Gibson, C. (2014). Music festivals and regional development policy: Towards a festival ecology. *Perfect Beat*, *14*(2), 140–57. http://dx.doi.org/10.1558/prbt. v14i2.140

Gibson, C., Connell, J., Waitt, G., & Walmsley, J. (2011). The extent and significance of rural festivals. In C. Gibson and J. Connell (Eds.), *Festival Places: Revitalising Rural Australia*, 3–24. Bristol, UK: Channel View Publications.

Gibson, C., & Dunbar-Hall, P. (2006). Nitmiluk: Place, politics and empowerment in Australian Aboriginal popular music. In J.C. Post (Ed.), *Ethnomusicology: A Contemporary Reader*, 383–400. New York: Taylor & Francis.

Gibson, C., Waitt, G., Walmsley, J., & Connell, J. (2010). Cultural festivals and economic development in nonmetropolitan Australia. *Journal of Planning Education and Research*, *29*(3), 280–293. http://dx.doi. org/10.1177/0739456X09354382

Gibson-Wood, H., & Wakefield, S. (2013). "Participation," white privilege and environmental justice: Understanding environmentalism among Hispanics in Toronto. *Antipode*, *45*(3), 641–62. http://dx.doi.org/10.1111/j.1467-8330.2012.01019.x

Gieseking, J.J. & Mangold, W. (2014a). Place and identity. In J.J. Gieseking & W. Mangold (Eds.), *The People, Place, and Space Reader*, 73–81. New York: Routledge.

Gieseking, J.J. & Mangold, W. (2014b). Diverse conceptions between people, place and space. In J.J. Gieseking & W. Mangold (Eds.), *The People, Place, and Space Reader*, 3–6. New York: Routledge.

Gilbert, L. (2014). Social Justice and the" Green" City. *Revista Brasileira de Gestão Urbana*, *6*(2), 158–69. http://dx.doi.org/10.7213/urbe.06.002.SE01

Gillberg, N., & Adolfsson, P. (2014). *Proud to Be Pride: A Discourse Analysis of the Presentation of Diversity on City Websites*. Gothenburg: Gothenburg Research Institute.

Gilmartin, M. (2008). Migration, identity and belonging. *Geography Compass*, *2*(6), 1837–52. http://dx.doi.org/10.1111/j.1749-8198.2008.00162.x

Gilmore, A. (2014). Evaluating legacies: Research, evidence and the regional impact of the Cultural Olympiad. *Cultural Trends*, *23*(1), 29–41. http://dx.doi.org/10.1080/09548963.2014.862001

Gilster, M.E. (2014). Putting activism in its place: The neighborhood context of participation in neighborhood-focused activism. *Journal of Urban Affairs*, *36*(1), 33–50. http://dx.doi.org/10.1111/juaf.12013

Ginieniewicz, J. (2010). Identity politics and political representation of immigrants: The perceptions of Latin Americans in Toronto. *Journal of*

Immigrant & Refugee Studies, 8(3), 261–83. http://dx.doi.org/10.1080/15562948.2010.501280

Ginwright, S., Noguera, P., & Cammarota, J. (Eds.). (2013). *Beyond Resistance: Youth Activism and Community Change.* New York: Routledge.

Ginsburg, M., & Craig, A. (2010). Tradition becomes the teacher: Community events enrich educators' professional learning. *Journal of Staff Development, 31*(4), 36–41.

Gioia, D. (2014). Reflections on teaching ethnographic fieldwork: Building community participatory practice. *Qualitative Social Work, 13*(1), 144–53. http://dx.doi.org/10.1177/1473325013509301

Giorgi, L. Sassatelli, M., & Delanty, G. (Eds.). (2011). *Festivals and the Cultural Public Sphere.* London: Routledge.

Girit Heck, O. (2011). *Representing Turkish National Culture and Turkish-American Identity in Chicago's Turkish Festivals.* Doctoral dissertation, University of Iowa. http://ir.uiowa.edu/etd/2502.

Godfrey, B.J., & Arguinzoni, O.M. (2012). Regulating public space on the beachfronts of Rio de Janeiro. *Geographical Review, 102*(1), 17–34. http://dx.doi.org/10.1111/j.1931-0846.2012.00128.x

Goh, D.P.S. (2011). State carnivals and the subvention of multiculturalism in Singapore. *British Journal of Sociology, 62*(1), 111–33. http://dx.doi.org/10.1111/j.1468-4446.2010.01347.x

Goh, D.P.S. (2013). Multicultural carnivals and the politics of the spectacle in global Singapore. *Inter-Asia Cultural Studies, 14*(2), 228–51. http://dx.doi.org/10.1080/14649373.2013.769751

Goldblatt, J.J. (1997). *Special Events: Best Practices in Event Management.* 2nd ed. New York: YNR.

Goldblatt, J.J. (2002). *Special Events: Twenty-first Century Global Event Management.* London: Wiley.

Gongaware, T.B. (2010). Collective memory anchors: Collective identity and continuity in social movements. *Sociological Focus, 43*(3), 214–39. http://dx.doi.org/10.1080/00380237.2010.10571377

Gonzalez, J.J. (2013). *A Model for Booth Assignments during Fiesta.* San Antonio: University of Texas, College of Business.

Goonewardena, K., & Kipfer, S. (2005). Spaces of difference: Reflections from Toronto on multiculturalism, bourgeois urbanism and the possibility of radical urban politics. *International Journal of Urban and Regional Research, 29*(3), 670–8. http://dx.doi.org/10.1111/j.1468-2427.2005.00611.x

Gordon, S., Adler, H., & Scott-Halsell, S. (2014). Career fairs: Are they valuable events? Hospitality and tourism recruiter perceptions of attributes towards participation and activities. *International Journal of Hospitality and Event Management, 1*(1), 81–94. http://dx.doi.org/10.1504/IJHEM.2014.062858

Gotham, K.F. (2005). Theorizing urban spectacles. *City, 9*(2), 225–46.
Gough, G., & Longhurst, J., and the University of the West of England. (2014). Raising sustainability awareness: Engaging the whole institutional community. Paper presented at the conference "*On Sustainability: Tenth International Conference on Environmental, Cultural, Economic and Social Sustainability*," 22–4 Jan., Split, Croatia.
Gouzouasis, P., & Henderson, A. (2012). Secondary student perspectives on musical and educational outcomes from participation in band festivals. *Music Education Research, 14*(4), 479–98. http://dx.doi.org/10.1080/14613808.2012.714361
Graff, A. (2010). Looking at pictures of gay men: Political uses of homophobia in contemporary Poland. *Public Culture, 22*(3), 583–603. http://dx.doi.org/10.1215/08992363-2010-010
Graham, R.S. (2013). *"Tell Me What You Eat and I Will Tell You Who You Are (and Where You Are From)": Food, Culture and Re-Membering In New Zealand, a Case Study Approach.* Doctoral dissertation, University of Waikato.
Grames, E., & Vitcenda, M. (2012). Community festivals – big benefits, but risks, too. University of Minnesota Extension. *Community Features* (Winter). http://www.extension.umn.edu/community/news/community-festivals/
Grams, D.M. (2013). Freedom and cultural consciousness: Black working-class parades in post-Katrina New Orleans. *Journal of Urban Affairs, 35*(5), 501–29. http://dx.doi.org/10.1111/juaf.12026
Grazian, D. (2010). Demystifying Authenticity in the Sociology of Culture. In J.R. Hall & M. Grindstaff (Eds.), *Handbook of Cultural Sociology,* 191–200. New York: Routledge.
Green, G.P., & Haines, A. (2011). *Asset Building and Community Development.* Thousand Oaks, CA: Sage.
Greene, S.J. (2003). Staged cities: Mega-events, slum clearance, and global capital. *Yale Human Rights and Development Law Journal, 6*(1), 61.
Greenhaigh, P. (2011). *Fair World: A History of World's Fairs and Expositions, from London to Shanghai, 1851–2010.* Winterbourne, UK: Papadakis.
Greensmith, C., & Giwa, S. (2013). Challenging settler colonialism in contemporary queer politics: Settler homonationalism, pride Toronto, and two-spirit subjectivities. *American Indian Culture and Research Journal, 37*(2), 129–48.
Grippo, R.M., & Hoskins, C. (2004). *Macy's Thanksgiving Day Parade (NY).* New York: Academic.
Grodach, C. (2010). Art spaces, public space, and link to community development. *Community Development Journal: An International Forum, 45*(4), 474–93. http://dx.doi.org/10.1093/cdj/bsp018

Grunwell, S., Ha, I., & Swanger, S.L. (2011). Evaluating the economic and fiscal impact of an international cultural heritage festival on a regional economy: Folkmoot USA. *Tourism Culture & Communication, 11*(2), 117–30. http://dx.doi.org/10.3727/109830411X13215686205969

Guazon, T.M. (2013). Creative mediations of the city: Contemporary public art as compass of metro Manila's urban conditions. *International Journal of Urban and Regional Research, 37*(3), 864–78. http://dx.doi.org/10.1111/j.1468-2427.2013.01211.x

Gudelunas, D. (2011). Consumer myths and the gay men and women who believe them: A qualitative look at movements and markets. *Psychology and Marketing, 28*(1), 53–68. http://dx.doi.org/10.1002/mar.20380

Gundlach, H., & Neville, B. (2011). Authenticity: Further theoretical and practical development. *Journal of Brand Management, 19*(4), 484–99.

Gunsoy, E., & Hannam, K. (2013). Festivals, community development and sustainable tourism in the Karpaz region of Northern Cyprus. *Journal of Policy Research in Tourism, Leisure and Events, 5*(1), 81–94. http://dx.doi.org/10.1080/19407963.2013.774201

Gursoy, D., Kim, K., & Uysal, M. (2004). Perceived impacts of festivals and special events by organizers: An extension and validation. *Tourism Management, 25*(2), 171–81. http://dx.doi.org/10.1016/S0261-5177(03)00092-X

Guttormsen, T., & Fageraas, K. (2011). The social production of "attractive authenticity" at the World Heritage Site of Roros, Norway. *International Journal of Heritage Studies, 17*(5), 442–62. http://dx.doi.org/10.1080/13527258.2011.571270

Hackworth, J., & Rekers, J. (2005). Ethnic packaging and gentrification: The case of four neighborhoods in Toronto. *Urban Affairs Review, 41*(2), 211–36. http://dx.doi.org/10.1177/1078087405280859

Haines, A. (2009). Asset-based community development. In R. Phillips & R.H. Pitman (Eds.), *An Introduction to Community Development*, 38–48. New York: Routledge.

Hains, D. (2014, 23 June). Toronto's Pride Festival evolves into economic powerhouse. *Globe and Mail*, B2.

Hampton, E., & Licona, M. (2013). Examining the Impact of Science Fairs in a Mexican-American Community. *Journal of Border Educational Research, 5*(1).

Han, K., Shih, P.C., Rosson, M.B., & Carroll, J.M. (2014). Enhancing community awareness of participation in local heritage with a mobile application. In *Proceedings of the 17th ACM Conference on Computer Supported Cooperative Work and Social Computing*, 1144–55. New York: ACM. http://dx.doi.org/10.1145/2531602.2531640

Hansen, B.C. (2014). Oral history project: Lithuanian students study the past to gain skills for the future. *International Journal of Education*, *6*(3), 216–28. http://dx.doi.org/10.5296/ije.v6i3.6229

Hanson, J. (2011). *Inside the Body Politic: Examining the Birth of Gay Liberation*. Thesis, Ohio State University.

Harney, N.D. (2006). The politics of urban space: Modes of place-making by Italians in Toronto's neighbourhoods. *Modern Italy*, *11*(1), 25–42. http://dx.doi.org/10.1080/13532940500489544

Harnik, P., & Crompton, J.L. (2014). Measuring the total economic value of a park system to a community. *Managing Leisure*, *19*(3), 188–211.

Hart, R.A. (2013). *Children's Participation: The Theory and Practice of Involving Young Citizens in Community Development and Environmental Care*. New York: Routledge.

Hart, R.A. (2014). Containing children: Some lessons on planning for play from New York City. In J.J. Gieseking & W. Mangold (Eds.), *The People, Place, and Space Reader*, 416–20. New York: Routledge.

Hartel, R.W., & Hartel, A. (2014). Cotton candy. In *Candy Bites: The Science of Sweets*, 37–40. Springer New York: Springer. http://dx.doi.org/10.1007/978-1-4614-9383-9_10

Harvey, D. (2009). *Social Justice and the City*. Rev. ed. Athens, GA.: University of Georgia Press.

Hass, K.A. (1998). *Carried to the Wall: American Memory and the Vietnam Veterans Memorial*. Los Angeles: University of California Press.

Hatfield, E. (2013, 6 June). Toronto's Portugal Day has a long history: Introduced to Portugal in 1910 as a way to secularize the country. *Parkdale Villager*. http://www.insidetoronto.com/news-story/3411603-toronto-s-portugal-day-has-a-long-history/

Hatton, T.J., & Leigh, A. (2011). Immigrants assimilate as communities, not just as individuals. *Journal of Population Economics*, *24*(2), 389–419. http://dx.doi.org/10.1007/s00148-009-0277-0

Hawdon, J., & Ryan, J. (2011). Social relations that generate and sustain solidarity after a mass tragedy. *Social Forces*, *89*(4), 1363–84. http://dx.doi.org/10.1093/sf/89.4.1363

Hawkins, E.R., & Brice, J.H. (2010). Fire jumpers: Description of burns and traumatic injuries from a spontaneous mass gathering and celebratory riot. *Journal of Emergency Medicine*, *36*(2), 162–7.

Hayden, D. (1995). *The Power of Place: Urban Landscapes as Public History*. Cambridge, MA: MIT Press.

Hayes-Bautista, D.E. (2012). *El Cinco de Mayo: An American Tradition*. Los Angeles: University of California Press.

Heitmann, S. (2011). Authenticity in tourism. In P. Robertson, S. Heitmann, & P. Dieke (Eds.), *Research Themes for Tourism*, 45–58. Boston: CBI.

Helbing, D., & Mukerji, P. (2012). Crowd disasters as systemic failures: Analysis of the Love Parade disaster. *EPJ Data Science*, *1*(1), 1–40.

Henry, R., & Foana'ota, L. (2015). Heritage transactions at the Festival of Pacific Arts. *International Journal of Heritage Studies*, *21*(2), 133–52.

Herzog, L.A. (2004). Globalization of the barrio: Transformation of the Latino cultural landscapes of San Diego, California. In D.B. Arreola (Ed.), *Hispanic spaces, Latino spaces: Community and cultural diversity in contemporary America*, 103–24. Austin: University of Texas Press.

Heywood, P. (2011). *Community Planning: Integrating Social and Physical Environment*. New York: Wiley.

Higgins, L. (2012). *Community Music: In Theory and in Practice*. New York: Oxford University Press. http://dx.doi.org/10.1093/acprof:oso/9780199777839.001.0001

Higgs, D., & Anderson, G.M. (2013). Portuguese: Portuguese explorers were among the first Europeans to see Canadian soil. Historica Canada. http://www.thecanadianencyclopedia.ca/en/article/portuguese/

Hill, G. (2014). Production of heritage: The Basque block in Boise, Idaho. *BOGA: Basque Studies Consortium Journal*, *1*(2), art3.

Hoefferle, M.M. (2012). Floats, friendship and fun: Exploring motivations for community art engagement. *International Journal of Education Through Art*, *8*(3), 253–69. http://dx.doi.org/10.1386/eta.8.3.253_1

Hogan, T., Bunnell, T., Pow, C.P., Permanasari, E., & Morshidi, S. (2012). Asian urbanisms and the privatization of cities. *Cities*, *29*(1), 59–63. http://dx.doi.org/10.1016/j.cities.2011.01.001

Holloway, I., Brown, L., & Shipway, R. (2010). Meaning not measurement: Using ethnography to bring a deeper understanding to the participant experience of festivals and events. *International Journal of Event and Festival Management*, *1*(1), 74–85. http://dx.doi.org/10.1108/17852951011029315

Hollows, J., Jones, S., Taylor, B., & Dowthwaite, K. (2013). Making sense of urban food festivals: Cultural regeneration, disorder and hospitable cities. *Journal of Policy Research in Tourism, Leisure and Events*, *6*(1), 1–14.

Holyfield, L., Cobb, M., Murray, K., & McKinzie, A. (2013). Musical ties that bind: Nostalgia, affect, and heritage in festival narratives. *Symbolic Interaction*, *36*(4), 457–77. http://dx.doi.org/10.1002/symb.67

Hoobler, D., & T. Hoobler. (2012). *From Street Fair to Medical Home: Charles B. Wang Community Health Center – Chinatown Health Clinic*. New York: Charles B. Wang Community Health Center.

Hood, L. (2007). Radio reverb: The impact of "local" news reimported to its own community. *Journal of Broadcasting & Electronic Media, 51*(1), 1–19. http://dx.doi.org/10.1080/08838150701307970

Hope, D.P. (2010). From the stage to the grave: Exploring celebrity funerals in dancehall culture. *International Journal of Cultural Studies, 13*(3), 254–70. http://dx.doi.org/10.1177/1367877909359733

Hopkins, R. (2011). *The Transition Companion: Making Your Community More Resilient in Uncertain Times.* White River Junction, VT: Chelsea Green Publishing.

Horng, J.S., Monica Hu, M.L., Teng, C.C., Hsiao, H.L., Tsai, C.Y., & Liu, C.H. (2014). How the introduction of concepts of energy saving and carbon reduction (ESCR) can affect festival visitors' behavioural intentions: An investigation using a structural model. *Journal of Sustainable Tourism, 22*(8), 1216–35.

Hou, J. (Ed.). (2013). *Transcultural Cities: Border-Crossing and Placemaking.* New York: Routledge.

Howley, K. (2005). *Community Media: People, Places, and Communication Technologies.* Cambridge: Cambridge University Press.

Hoxsey, D. (2012). *Whose Pride?: An Institutional Ethnography on Participating in Toronto's Pride Parade.* Doctoral dissertation, University of Victoria.

Huang, J.Z., Li, M., & Cai, L.A. (2010). A model of community-based festival image. *International Journal of Hospitality Management, 29*(2), 254–60. http://dx.doi.org/10.1016/j.ijhm.2009.10.010

Huang, L., & Zhao, W.M. (2013). Cultural planning for urban spaces: Cultural turn of contemporary urban planning. *Advanced Materials Research, 790,* 492–6. http://dx.doi.org/10.4028/www.scientific.net/AMR.790.492

Huang, L.-L., & Hsu, J.-Y. (2011). From cultural building, economic revitalization to local partnership? The changing nature of community mobilization in Taiwan. *International Planning Studies, 16*(2), 131–50. http://dx.doi.org/10.1080/13563475.2011.561058

Hudes, S. (2014, 24 June). Anti-Israel group to march in Toronto Pride parade, but with little opposition this year. *National Post,* 1.

Hulbin, X., & Marzuki, A. (2012). Community participation of cultural heritage tourism from innovation system perspective. *International Journal of Services Technology & Management, 16*(3), 106–27.

Hum, T. (2004). Asian immigrant settlements in New York City: Defining "communities of interest." *AAPI Nexus: Policy, Practice and Community, 2*(2), 20–48.

Hutchins, F.T., Brown, L.D., & Poulsen, K.P. (2014). An anthropological approach to teaching health sciences students cultural competency in a field school program. *Academic Medicine, 89*(2), 251–6. http://dx.doi.org/10.1097/ACM.0000000000000088

Hutchison, P., & Lord, J. (2012). Community-based research and leisure scholarship: A discernment process. *Leisure/Loisir, 36*(1), 65–83.

Hutton, A., Brown, S., & Verdonk, N. (2013). Exploring culture: Audience predispositions and consequent effects on audience behavior in a mass-gathering setting. *Prehospital and Disaster Medicine, 28*(3), 292–7. http://dx.doi.org/10.1017/S1049023X13000228

Hutton, A., Zeitz, K., Brown, S., & Arbon, P. (2012). Assessing the psychological elements of crowds and mass gatherings. *Prehospital and Disaster Medicine, 26*(6), 414–21. http://dx.doi.org/10.1017/S1049023X12000155

Hvenegaard, G.T. (2011). Potential conservation of wildlife festivals. *Event Management, 15*(4), 373–86. http://dx.doi.org/10.3727/152599511X13175676722645

Hvengaard, G.T., Delamere, T.A., Lemelin, R.H., Breager, K., & Anger, A. (2013). Insect festivals: Celebrating and fostering human-insects interactions. In R.H. Lemelin (Ed.), *The Management of Insects in Recreation and Tourism,* 198–216). New York: Cambridge University Press.

Hwang, D., Stewart, W.P., & Ko, D.-W. (2012). Community behavior and sustainable rural tourism development. *Journal of Travel Research, 51*(3), 328–41. http://dx.doi.org/10.1177/0047287511410350

Hyde, C.A., & Meyer, M. (2010). Does attitudinal consensus lead to neighborhood collaboration? An exploratory analysis of an urban community. *Journal of Human Behavior in the Social Environment, 20*(1), 56–73. http://dx.doi.org/10.1080/10911350903126932

Ibrahim, S., & Sidani, S. (2014). Strategies to recruit minority persons: A systematic review. *Journal of Immigrant and Minority Health, 16*(5), 882–8.

Ikuomola, A.D., Okunola, R.A., & Akindutire, A.F. (2014). Ritualised (Dis) order: Street carnivals, transgression and excesses in Nigeria. *Antropologija, 14*(1), 129–46. http://www.anthroserbia.org/Content/PDF/Articles/b68d3c9ff8f345a0b320b45419696e17.pdf

Ingalls, M.M. (2012). Singing praise in the streets: Performing Canadian Christianity through public worship in Toronto's Jesus in the City parade. *Culture and Religion, 13*(3), 337–59. http://dx.doi.org/10.1080/14755610.2012.706230

Isaac-Flavien, J. (2013). The translation of carnival in Trinidad and Tobago: The evolution of a festival. *Tusaaji: A Translation Review, 2*(2), 42–55.

Idoko, A.A. (2012). Culture and empowerment: An appraisal of Oje'nmeho Cultural Festival of Awume. *IAMURE Multidisciplinary Research, 3*(1).

Illiyas, F.T., Mani, S.K., Pradeepkumar, A.P., & Mohan, K. (2013). Human stampedes during religious festivals: A comparative review of mass gathering emergencies in India. *International Journal of Disaster Risk Reduction, 5*(1), 10–18. http://dx.doi.org/10.1016/j.ijdrr.2013.09.003

Iltis, A. (2012). Ritual as the creation of social reality. In D. Solomon, R. Fan, & P.-C. Lu (Eds.), *Ritual and the Moral Life*, 17–28. New York: Springer.

Inglis, C. (2011). Chinatown Sydney: A window on the Chinese community. *Journal of Chinese Overseas*, 7(1), 45–68. http://dx.doi.org/10.1163/179325411X565407

Iordanova, D. & Torchin, L. (Eds.). (2012). *Film Festivals and Activism*. St Andrews, Scotland: St Andrews Film Studies.

Isoldi, K.K., Dalton, S., Rodriguez, D.P., & Nestle, M. (2012). Classroom "cupcake" celebrations: Observations of foods offered and consumed. *Journal of Nutrition Education and Behavior*, 44(1), 71–5. http://dx.doi.org/10.1016/j.jneb.2011.03.144

Israel, B. A., Schulz, A. J., Parker, E. A., Becker, A. B., Allen III, A. J., & Guzman, J. R. (2008). Critical issues in developing and following CBPR principles. In M. Minkler, & N. Wallerstein (Eds.), *Community-Based Participatory Research for Health: From Process to Outcomes*, 2nd ed., 47–66. San Francisco: Jossey-Bass.

Jackson, I. (2008). *Celebrating Communities: Community Festivals, Participation and Belonging*. University of Melbourne: Australian Centre, School of Historical Studies. UNESCO e-Journal, 161–72.

Jackson, S.F., Cleverly, S., Poland, B., Burman, D., Edwards, R. & Robertson, A. (2003). Working with Toronto neighborhoods towards developing indicators of community capacity. *Health Promotion International*, 18(4), 339–350. http://dx.doi.org/10.1093/heapro/dag415

Jaeger, K., & Mykletun, R.J. (2013). Festivals, identities, and belonging. *Event Management*, 17(3), 213–26. http://dx.doi.org/10.3727/152599513X13708863377791

Jaguaribe, B. (2013a). Imagens da multitao: Carnaval e midia. *E-Compas*, 16(3).

Jaguaribe, B. (2013b). Carnival crowds. *Sociological Review*, 61(1), 69–88. http://dx.doi.org/10.1111/1467-954X.12054

Jaimangal-Jones, D. (2014). Utilising ethnography and participant observation in festival and event research. *International Journal of Event and Festival Management*, 5(1), 39–55. http://dx.doi.org/10.1108/IJEFM-09-2012-0030

Jakob, D. (2013). The eventification of place: Urban development and experience consumption in Berlin and New York City. *European Urban and Regional Studies*, 20(4), 447–59. http://dx.doi.org/10.1177/0969776412459860

Jamieson, K. (2013). Tracing festival imaginaries: Between affective urban idioms and administrative assemblages. *International Journal of Cultural Studies*, 17(3), 293–303.

Jamir, L., Nongkynrih, B., & Gupta, S.K. (2014). Community noise pollution in urban India: Need for public health action. *Indian Journal of Community Medicine*, 39(1), 8–12. http://dx.doi.org/10.4103/0970-0218.126342

Janiskee, R.L. (1995). The temporal distribution of America's community festivals. *Festival Management and Event Tourism, 3*(3), 129–37.

Jarman, N. (2003). From outrage to apathy? The disputes over parades, 1995–2003. *Global Review of Ethnopolitics, 3*(1), 92–105. http://dx.doi.org/10.1080/14718800308405160

Jayaswal, T. (2010). *Events-Induced Tourism: A Protocol Analysis.* Doctoral dissertation, Auckland University of Technology.

Jeffres, L.W. (2010). The communicative city: Conceptualizing, operationalizing, and policy making. *Journal of Planning Literature, 25*(2), 99–110. http://dx.doi.org/10.1177/0885412210369455

Jepson, A., Clarke, A., & Ragsdell, G. (2014). Investigating the application of the motivation-opportunity-ability model to reveal factors which facilitate or inhibit inclusive engagement within local community festivals. *Scandinavian Journal of Hospitality and Tourism, 14*(3), 331–48.

Jepson, A., Wiltshier, P., & Clarke, A. (2008). *Community Festivals: Involvement and Inclusion.* Northampton: University of Northampton Press.

Jiang, J., & Schmader, S.W. (2014). Event management education and professionalism: The view from the trenches. *Event Management, 18*(1), 25–37. http://dx.doi.org/10.3727/152599514X13883555341814

Jiménez, K.P. (2008). Latina Landscape: Queer Toronto. *Canadian Journal of Environmental Education, 13*(2), 114–29.

Johansson, M., & Kociatkiewicz, J. (2011). City festivals: Creativity and control in staged urban experiences. *European Urban and Regional Studies, 18*(4), 392–405. http://dx.doi.org/10.1177/0969776411407810

Johnston, M.E., Twynam, G.D., & Farrell, J.M. (1999). Motivation and satisfaction of event volunteers for a major youth organization. *Leisure/Loisir, 24*(1–2), 161–77.

Jones, E. (2014). Multicultural screening and assessment of immigrants & refugees. Paper presented at the 142nd APHA Annual Meeting and Exposition, 15–19 Nov., New Orleans.

Jones, M.L. (2014). *Sustainable Event Management: A Practical Guide.* New York: Routledge.

Jorae, W.R. (2009). *The Children of Chinatown: Growing Up Chinese American in San Francisco, 1850–1920.* Chapel Hill, NC: University of North Carolina Press. http://dx.doi.org/10.5149/9780807898581_jorae

Jordan, E.J. (2015). Planning as a coping response to proposed tourism development. *Journal of Travel Research, 54*(3), 316–28.

Jukova, S. (2014). The role of ethno-cultural organizations in immigrant integration: A case study of the Bulgarian Society of Western Canada. *Canadian Ethnic Studies, 46*(1), 23–44.

Kallen, E. (1996). Gay and Lesbian rights issues: A comparative analysis of Sydney, Australia, and Toronto, Canada. *Human Rights Quarterly, 18*(1), 206–23. http://dx.doi.org/10.1353/hrq.1996.0008

Kallio, K.P., & Hakti, J. (2011). Young people's voiceless politics in the struggle over urban space. *GeoJournal, 76*(1), 63–75. http://dx.doi.org/10.1007/s10708-010-9402-6

Kaplan, D.H. & Li, W. (Eds.). (2006). *Landscapes of the Ethnic Economy.* Lantham, MD: Rowman & Littlefield.

Kates, S.M., & Belk, R.W. (2001). The meanings of lesbian and gay pride day resistance through consumption and resistance to consumption. *Journal of Contemporary Ethnography, 30*(4), 392–429. http://dx.doi.org/10.1177/089124101030004003

Kaufman, D. (2011). *Exploring Authenticity in American Indian Art at the First Festival.* Dissertation, George Mason University.

Kaufman, S., Ozawa, C.P., & Shmueli, D.F. (2014). Evaluating participatory decision processes: Which methods inform reflective practice? *Evaluation and Program Planning, 42*(1), 11–20. http://dx.doi.org/10.1016/j.evalprogplan.2013.08.002

Kazin, A. (2008). Fear of the city, 1783–1983. In F. Siegel & H. Siegel (Eds.), *Urban Society,* 13th ed., 2–7. Dubuque, IA: Contemporary Learning Series.

Kelly, C. (2013). Towards renewed descriptions of Canadian disability movements: Disability activism outside of the non-profit sector. *Canadian Journal of Disability Studies, 2*(1), 1–27. http://dx.doi.org/10.15353/cjds.v2i1.68

Kendrick, V.L., & Haslam, R.A. (2010). The user experience of crowds: A human factors challenge. *Proceedings of the Human Factors and Ergonomics Society Annual Meeting, 54*(23), 2000–2004.

Kenny, A. (2011). Mapping the context: Insights and issues from local government development of music communities. *British Journal of Music Education, 28*(2), 213–26. http://dx.doi.org/10.1017/S0265051711000088

Kezer, Z. (2009). An imaginable community: The national culture of nation-building in early republic Turkey. *Society and Space, 27*(3), 508–30.

Khaire, M. (2012). Brian Moeran & Jesper Strandgaard Pederson (Eds.), *Negotiating Values in the Creative Industries: Fairs, Festivals, and Competitive Events.* [review] *Administrative Science Quarterly, 57*(3), 541–4.

Khan, V.S. (2013). Hindu festivals: Hazards to environment and ecology. *Journal of Indian Research, 1*(1), 136–41.

Kilanc, R. (2013). The importance of planned events based on cultural values on Turkish people in Amsterdam and Istanbul. *World Applied Sciences Journal, 27*(9), 1163–7.

Kim, N., Ahn, Y.J., & Wicks, B.E. (2014). Local festival quality and the application of a revised importance – performance analysis: The case of the Jirisan Cheon-Wang Festival. *Event Management*, *18*(2), 89–100. http://dx.doi.org/10.3727/152599514X13947236947266

Kim, S.E. (2013). *Experience and Perceived Value for Participants of Cultural and Art Festivals Organized for Persons with a Disability: A Korean Perspective*. Dissertation, Purdue University.

Kim, S.S., Choi, S., Agrusa, J., Wang, K.-C., & Kim, Y. (2010). The role of family decision makers in festival tourism. *International Journal of Hospitality Management*, *29*(2), 308–18. http://dx.doi.org/10.1016/j.ijhm.2009.10.004

Kim, Y.H., Duncan, J.L., & Jai, T.M. (2014). A case study of a southern food festival: Using a cluster analysis approach. *Anatolia: An International Journal of Tourism and Hospitality Research*, *25*(3), 457–73.

Kimmel, C.E., Hull, R.B., Stephenson, M.O., Robertson, D.P., & Cowgill, K.H. (2012). Building community capacity and social infrastructure through landcare: A case study of land grant engagement. *Higher Education*, *64*(2), 223–35. http://dx.doi.org/10.1007/s10734-011-9489-9

Knight, D. (2014). Dynamic health fair planning model for community partners provided by a local health department. Paper presented at the 142nd APHA Annual Meeting and Exposition, 15–19 Nov. 2014, New Orleans.

Kirk-Downey, T., & Perry, B. (2006). Making the transition to school a community event: The Wollongong experience. *International Journal of Transitions in Childhood*, *2*(1), 40–9.

Kitty K. (2014). Chinese New Year Parade. *Yelp*. http://www.yelp.com/biz/chinese-new-year-parade-san-francisco

Klein, R.R. (2011). Where music and knowledge meet: A comparison of temporary events in Los Angeles and Columbus, Ohio. *Area*, *43*(3), 320–6. http://dx.doi.org/10.1111/j.1475-4762.2011.00997.x

Klein, W., Köster, G., & Meister, A. (2010). Towards the calibration of pedestrian stream models. In R. Wyrzykowski, J. Dongarra, K. Karczewski, and J. Waśniewski, J. (Eds.), *Parallel Processing and Applied Mathematics*, 521–8. Berlin: Springer. http://dx.doi.org/10.1007/978-3-642-14403-5_55

Klemek, C. (2011). *The Transatlantic Collapse of Urban Renewal: Postwar Urbanism from New York to Berlin*. Chicago: University of Chicago Press.

Kline, M.V., & Huff, R.M. (2008). Tips for working with Asian American populations. In *Health Promotion in Multicultural Populations: A Handbook for Practitioners and Students*, 474–83. Los Angeles: Sage.

Knottnerus, J.D. (2010). Collective events, rituals, and emotions. *Advances in Group Processes*, *27*(1), 39–61. http://dx.doi.org/10.1108/S0882-6145(2010)0000027005

Koch, R., & Latham, A. (2012). Rethinking urban public space: Accounts from a junction in West London. *Transactions of the Institute of British Geographers, 37*(4), 515–29. http://dx.doi.org/10.1111/j.1475-5661.2011.00489.x

Koefoed, L., Simonsen, K., & de Neergaard, M. (2012). Paradoxical spaces: Encountering the other in public space. Project description at http://rucforsk.ruc.dk/site/files/48330989/FSE_projektbeskrivelse.pdf

Koeman, L., Kalnikaitė, V., Rogers, Y., & Bird, J. (2014). What chalk and tape can tell us: Lessons learnt for next generation urban displays. In *Proceedings of The International Symposium on Pervasive Displays*, 130. New York: ACM. http://dx.doi.org/10.1145/2611009.2611018

Kohl-Arenas, E., Nateras, M.M., & Taylor, J. (2014). Cultural organizing as critical praxis: Tamejavi builds immigrant voice, belonging, and power. *Journal of Poverty, 18*(1), 5–24. http://dx.doi.org/10.1080/10875549.2013.866804

Kolomer, S., Quinn, M.E., & Steele, K. (2010). Interdisciplinary health fairs for older adults and the value of interprofessional service learning. *Journal of Community Practice, 18*(2–3), 267–79. http://dx.doi.org/10.1080/10705422.2010.485863

Kondakov, A. (2013). Resisting the salience: The use of tolerance and equality arguments by gay and lesbian activist groups in Russia. *Canadian Journal of Law and Society/Revue canadienne droit social, 28*(3), 403–24.

Kong, S. (2013). Space of possibilities: Civic discourse and multicultural citizenship in locally produced Chinese television programs in Metro Vancouver. *Journal of International Migration and Integration, 14*(1), 119–38.

Koopman, S. (2013). A liberatory space? Rumors of rapes at the 5th World Social Forum, Porto Alegre, 2005. *Journal of International Women's Studies, 8*(3), 149–63.

Köster, M. (2014). Brazilian brokers, boundaries and buildings: A material culture of politics. *Journal of Material Culture, 19*(2), 125–44.

Kostopoulou, S., & Kalogirou, S. (2011). The spatial-economic impact of cultural events. *International Journal of Sustainable Development, 14*(3–4), 309–31. http://dx.doi.org/10.1504/IJSD.2011.041967

Kostopoulou, S., Vagionis, N., & Kourkouridis, D. (2013). Cultural festivals and regional economic development: Perceptions of key interest groups. In A. Matias, P. Nijkamp, & M. Sarmento (Eds.), *Quantitative Methods in Tourism Economics*, 175–94. New York: Springer.

Kotin, S., Dyrness, G.R., & Irazábal, C. (2011). Immigration and integration: Religious and political activism for/with immigrants in Los Angeles. *Progress in Development Studies, 11*(4), 263–84. http://dx.doi.org/10.1177/146499341001100401

Koutrolikou, P.P. (2012). Spatialities of ethnocultural relations in multicultural East London: Discourses of interaction and social mix. *Urban Studies (Edinburgh, Scotland), 49*(10), 2049–66. http://dx.doi.org/10.1177/0042098011422569

Kretzmann, J.P., & McKnight, J. (1993). *Building Communities from the Inside Out: A Path toward Finding and Mobilizing a Community's Assets.* Evanston, IL: Center for Urban Affairs and Policy Research, Northwestern University.

Kretzmann, J.P., & McKnight, J. (1996). *A Guide to Mapping and Mobilizing the Economic Capacities of Local Residents.* Evanston, IL: Center for Urban Affairs and Policy Research, Northwestern University.

Kroeger, J. (2014). Where do we go from here in family engagement? Comments at the late turn of the 21st century. *Journal of Family Diversity in Education, 1*(1).

Kuecker, G., Mulligan, M., & Nadarajah, Y. (2011). Turning to community in times of crisis: Globally derived insights on local community formation. *Community Development Journal: An International Forum, 46*(2), 245–64. http://dx.doi.org/10.1093/cdj/bsq002

Kuhar, R., & Svab, A. (2014). The only gay in the village? Everyday life of gays and lesbians in rural Slovenia. *Journal of Homosexuality, 61*(8), 1091–16.

Kurdia, D. (2013). *Who Produces Urban Space? Gentrification and Contestations over Urban "Authenticity."* Dissertation, University of Western Ontario. Thesis and Dissertation Repository 1405.

Kuroishi, I. (2013). Shinjo Matsuri: Community revitalization through the construction and circulation of decorated floats. *Paragrana, 22*(1), 180–94. http://dx.doi.org/10.1524/para.2013.0013

Lacroix, T., & Fiddian-Qasmiyeh, E. (2013). Refugee and diaspora memories: The Politics of remembering and forgetting. *Journal of Intercultural Studies, 34*(6), 684–96. http://dx.doi.org/10.1080/07256868.2013.846893

Lade, C., & Jackson, J. (2004). Key success factors in regional festivals: Some Australian experiences. *Event Management, 9*(1), 1–11. http://dx.doi.org/10.3727/1525995042781066

Laflin, S.L., & Anderson, D.E. (2010). State fairs: A means of exposing America's youth to the veterinary profession? *Online Journal of Rural Research & Policy, 5*(7), 13. http://dx.doi.org/10.4148/ojrrp.v5i7.270

LaFrance, J., Nichols, R., & Kirkhart, K.E. (2012). Culture writes the script: On the centrality of context in indigenous evaluation. *New Directions for Evaluation, 2012*(135), 59–74. http://dx.doi.org/10.1002/ev.20027

Lahiri-Dutt, K., & Ahmad, N. (2012). In F. Vanclay and A.M. Esteves (Eds.), *Considering Gender in Social Impact Assessments: New Directions in Social Impact Assessments – Conceptual and Methodological Advances.*

Laing, J., & Frost, W. (2010). How green was my festival: Exploring challenges and opportunities associated with staging green events. *International Journal of Hospitality Management*, *29*(2), 261–7. http://dx.doi.org/10.1016/j.ijhm.2009.10.009

Lampel, J. (2011). Afterword: Converting values into other values – fairs and Festivals as resource valuation and trading events. In B. Moeran & J. Pedersen, (Eds.), *Negotiating Values in the Creative Industries: Fairs, Festivals and Competitive Events*, 334–47. New York: Cambridge University Press.

Landolt, P., & Goldring, L. (2009). Immigrant political socialization as bridging and boundary work: Mapping the multi-layered incorporation of Latin American immigrants in Toronto. *Ethnic and Racial Studies*, *32*(7), 1226–47. http://dx.doi.org/10.1080/01419870802604016

Landy, D.C., Gorin, M.A., & O'Connell, M.T. (2011). Student-led rural health fairs: Attempting to improve medical education and access to health care. *Southern Medical Journal*, *104*(8), 598–603. http://dx.doi.org/10.1097/SMJ.0b013e31822580a9

Langegger, S. (2013). Emergent public space: Sustaining Chicano culture in North Denver. *Cities*, *35*(1), 26–32. http://dx.doi.org/10.1016/j.cities.2013.04.013

Langhout, R.D. (2014). Questioning our questions: Assessing question asking practices to evaluate a yPAR program. *Urban Review*, *46*, 703–724.

Langstraat, F., & Van Melik, R. (2013). Challenging the "end of public space": A comparative analysis of publicness in British and Dutch urban spaces. *Journal of Urban Design*, *18*(3), 429–48.

Laing, J., Wheeler, F., Reeves, K., & Frost, W. (2014). Assessing the experiential value of heritage assets: A case study of a Chinese heritage precinct, Bendigo, Australia. *Tourism Management*, *40*(4), 180–92. http://dx.doi.org/10.1016/j.tourman.2013.06.004

Landers, F. (2012). Urban play: Imaginatively responsible behavior as an alternative to neoliberalism. *Arts in Psychotherapy*, *39*(3), 201–5. http://dx.doi.org/10.1016/j.aip.2011.12.006

Langen, F., & Garcia, B. (2009). *Measuring the impacts of large scale cultural events: A literature review*. Impacts 08. https://www.liv.ac.uk/media/livacuk/impacts08/pdf/pdf/Impacts08-FLangen_and_BGarcia_May_2009_Events_Review.pdf

Larsen, A.K. (2012). Village fairs and rural adventure tours in Norway. In H. Horakova and A. Boscoboinik (Eds.), *From Production to Consumption: Transformation of Rural Communities*, 131–48. Münster: LIT Verlag.

Larsen, L., Sherman, L.S., Cole, L.B., Karwat, D., Badiane, K., & Coseo, P. (2014). Social justice and sustainability in poor neighborhoods: Learning and living in Southwest Detroit. *Journal of Planning Education and Research*, *34*(1), 5–18. http://dx.doi.org/10.1177/0739456X13516498

Larson, M. (2011). Innovation and creativity in festival organizations. *Journal of Hospitality Marketing & Management, 20*(3–4), 287–310. http://dx.doi.org/10.1080/19368623.2011.562414

Lashley, C. & Morrison, A. (Eds.). (2013). *In Search of Hospitality.* New York: Routledge.

Lau, C.Y., & Li, Y. (2015). Producing a sense of meaningful place: Evidence from a cultural festival in Hong Kong. *Journal of Tourism and Cultural Change, 13*(1), 56–77.

Lauver, L.S. (2011). Health fairs as a forum to pique young children's interest in nursing. *Journal of Pediatric Nursing, 26*(4), 359–63. http://dx.doi.org/10.1016/j.pedn.2010.06.013

Lavery, S.H., Smith, M.L., Esparza, A.A., Hrushow, A., Moore, M., & Reed, D.F. (2005). The community action model: A community-driven model designed to address disparities in health. *American Journal of Public Health, 95*(4), 611–16. http://dx.doi.org/10.2105/AJPH.2004.047704

Lavrinec, J. (2013). Urban scenography: Emotional and bodily experience. *Limes: Borderland Studies, 6*(1), 21–31.

Lawton, L.J., & Weaver, D.B. (2010). Normative and innovative sustainable resource management at birding festivals. *Tourism Management, 31*(4), 527–36. http://dx.doi.org/10.1016/j.tourman.2009.06.004

Le, H., Polonsky, M., & Arambewela, R. (2015). Social inclusion through cultural engagement among ethnic communities. *Journal of Hospitality Marketing & Management, 24*(4), 375–400.

Leal, J. (2014). What's (not) in a parade? Nationhood, ethnicity and regionalism in a diasporic context. *Nations and Nationalism, 20*(2), 200–17. http://dx.doi.org/10.1111/nana.12062

Lee, I.S., & Arcodia, C. (2011). The role of regional food festivals for destination branding. *International Journal of Tourism Research, 13*(4), 355–67. http://dx.doi.org/10.1002/jtr.852

Lee, I.S., Arcodia, C., & Lee, T.J. (2012). Key characteristics of multicultural festivals: A critical review of the literature. *Event Management, 16*(1), 93–101. http://dx.doi.org/10.3727/152599512X13264729827758

Lee, J. (2014). Visitors' emotional responses to the festival environment. *Journal of Travel & Tourism Marketing, 31*(1), 114–31. http://dx.doi.org/10.1080/10548408.2014.861726

Lee, J.E., Almanza, B.A., & Nelson, D.C. (2010). Food safety at fairs and festivals: Vendor knowledge and violations at a regional festival. *Event Management, 14*(3), 215–23. http://dx.doi.org/10.3727/152599510X12825895093632

Lee, J.J., & Kyle, G.T. (2014). Segmenting festival visitors using psychological commitment. *Journal of Travel Research*, *53*(5), 656–69.

Lee, J.J., Kyle, G., & Scott, D. (2012). The mediating effect of place attachment on the relationship between festival satisfaction and loyalty to the festival hosting destination. *Journal of Travel Research*, *51*(6), 754–67. http://dx.doi.org/10.1177/0047287512437859

Lee, J.-S., Lea, C.-K., & Choi, Y. (2010). Examining the role of emotional and functional values in festival evaluation. *Journal of Travel Research*, *50*(9), 685–96.

Lee, T.H. (2013). Influence analysis of community resident support for sustainable tourism. *Tourism Management*, *34*(1), 37–46. http://dx.doi.org/10.1016/j.tourman.2012.03.007

Lee, T.H., & Hsu, F.Y. (2013). Examining how attending motivation and satisfaction affects the loyalty for attendees at Aboriginal festivals. *International Journal of Tourism Research*, *15*(1), 18–34. http://dx.doi.org/10.1002/jtr.867

Lee, W. (2013). Whose festival is it? Patterns of participation in the Japan Matsuri. *Paragrana*, *22*(1), 165–79. http://dx.doi.org/10.1524/para.2013.0012

Leglar, M., & Smith, D.S. (2010). Community music in the United States: An overview of origins and evolution. *International Journal of Community Music*, *3*(3), 343–53. http://dx.doi.org/10.1386/ijcm.3.3.343_1

Leitão, F., Leitão, S.G., de Almeida, M.Z., Cantos, J., Coelho, T., & da Silva, P.E.A. (2012). Urban ethnobotany: Open-air fairs in the State of Rio de Janeiro, Brazil, as a potential source of new antitubercular plants. *Planta Medica*, *78*(11), PF41. http://dx.doi.org/10.1055/s-0032-1320588

Lekies, K.S. (2009). Youth engagement: The ethics of inclusion and exclusion. *Revue Les ateliers de l'éthique/The Ethics Forum*, *4*(1).

Leonard, M., & McKnight, M. (2013). Traditions and transitions: Teenagers' perceptions of parading in Belfast. *Children's Geographies*, *13*(4), 398–412.

Lena, J.C. (2011). Tradition and transformation at the Fan Fair Festival. In B. Moeran & J.S. Pedersen (Eds.), *Negotiating Values in the Creative Industries: Fairs, Festivals and Competitive Events*, 224–48. New York: Cambridge University Press. http://dx.doi.org/10.1017/CBO9780511790393.010

Lever, W.F. (2013). Evaluating the urban milieu of an individual city. In P.K. Kresl and J. Sobrino (Eds.), *Handbook of Research Methods and Applications in Urban Economies*, 372–95. Northampton, MA: Edward Elgar.

Lewis, H. (2010). Community moments: Integration of transnationism at "refugee" parties and events. *Journal of Refugee Studies*, *23*(4), 571–88. http://dx.doi.org/10.1093/jrs/feq037

Li, M. (2014). Ritual and social change: Chinese rural-urban migrant workers' spring festival homecoming as secular pilgrimage. *Journal of Intercultural*

Communication Research, 43(2), 113–33. http://dx.doi.org/10.1080/17475759. 2014.892896

Liberato, S.C., Brimblecombe, J., Ritchie, J., Ferguson, M., & Coveney, J. (2011). Measuring capacity building in communities: A review of the literature. *BMC Public Health, 11*(1), 850. http://dx.doi.org/10.1186/1471-2458-11-850

Lillywhite, J.M., Simonsen, J.E., & Wilson, B.J.H. (2013). A portrait of the US fair sector. *Event Management, 17*(3), 241–55. http://dx.doi.org/10.3727/152599513X13708863377872

Lim, M., & Barton, A.C. (2010). Exploring insideness in urban children's sense of place. *Journal of Environmental Psychology, 30*(3), 328–37. http://dx.doi.org/10.1016/j.jenvp.2010.03.002

Lin, J. (2011). *The Power of Urban Ethnic Places: Cultural Heritage and Community Life*. New York: Routledge.

Lindeman, S. (2014). "Until we live like they live in Europe": A multilevel framework for community empowerment in subsistence markets. *Journal of Macromarketing, 34*(2), 171–85.

Lindgren, A. (2013). The diverse city: Can you read all about it in ethnic newspapers? *Contemporary Readings in Law and Justice*, (Feb.), 120–40.

Lindholm, C. (2013). The rise of expressive authenticity. *Anthropological Quarterly, 86*(2), 361–95. http://dx.doi.org/10.1353/anq.2013.0020

Linko, M., & Silvanto, S. (2011). Infected by arts festivals: Federal policy and audience experiences in the Helsinki Metropolitan area. *Journal of Arts Management, Law, and Society, 41*(4), 224–39. http://dx.doi.org/10.1080/10632921.2011.624971

Litvin, S. (2013). Festivals and special events: Making the investment. *International Journal of Culture, Tourism and Hospitality Research, 7*(2), 184–7.

Litvin, S., Pan, B., & Smith, W. (2013). Festivals, special events, and the "rising tide." *International Journal of Culture, Tourism and Hospitality Research, 7*(2), 163–9.

Litman, T.L. (2012). *Community Cohesion as a Transport Planning Objective*. Victoria, BC: Victoria Policy Institute.

Little, R.M., & Froggett, L. (2010). Making meaning in muddy waters: Representing complexity through community-based storytelling. *Community Development Journal: An International Forum, 45*(4), 458–73. http://dx.doi.org/10.1093/cdj/bsp017

Lillywhite, J.M., Simonsen, J.E., & Acharya, R.N. (2013). Designing a better fair: How important are the animals? *Journal of Convention & Event Tourism, 14*(3), 217–35. http://dx.doi.org/10.1080/15470148.2013.810559

Littrell, J., Brooks, F., Ivery, J.M., & Ohmer, M.L. (2013). Inequality and its discontents. In L. Simmons & S. Harding (Eds.), *Economic Justice, Labor and Community Practice*, 1–30. New York: Routledge.

Liu, Y.-D., & Lin, C.-F. (2010). Critical factors for marketing urban cultural tourism, a conceptual model. *International Journal of Leisure and Tourism Marketing*, *1*(4), 379–91. http://dx.doi.org/10.1504/IJLTM.2010.032065

Livingston, K.A., Rosen, J.B., Zucker, J.R., & Zimmerman, C.M. (2014). Mumps vaccine effectiveness and risk factors for disease in households during an outbreak in New York City. *Vaccine*, *32*(3), 369–74. http://dx.doi.org/10.1016/j.vaccine.2013.11.021

Long, L. (2013). *Celebration of Community and Place: Who We Were, Who We Are, and What We Hope for the Future*. Auburn University Libraries. http://alamosindex.lib.auburn.edu/vufind/Record/AUpacers18

Loomis, B, Nguyen, Q. & Kim, A. (2011). Exposure to Pro-Tobacco Marketing and Promotions among New Yorkers. Albany, NY: New York State Department of Health. https://www.health.ny.gov/prevention/tobacco_control/docs/tobacco_marketing_exposure_rpt.pdf

Lopez Bonilla, J.M., Lopez-Bonilla, L.M., & Sanz-Altamira, B. (2010). Designated public festivals of interest to tourists. *European Planning Studies*, *18*(3), 435–47. http://dx.doi.org/10.1080/09654310903497728

Iorio, M., & Wall, G. (2012). Behind the masks: Tourism and community in Sardinia. *Tourism Management*, *33*(6), 1440–9. http://dx.doi.org/10.1016/j.tourman.2012.01.011

Loukaitou-Sideris, A. (2012). Addressing the challenges of urban landscapes: Normative goals for urban design. *Journal of Urban Design*, *17*(4), 467–84. http://dx.doi.org/10.1080/13574809.2012.706601

Loukaitou-Sideris, A., & Ehrenfeucht, R. (2009). *Sidewalks: Conflict and Negotiation over Public Space*. Cambridge, MA: MIT Press.

Loukaitou-Sideris, A., & Soureli, K. (2012). Cultural tourism as an economic development strategy for ethnic neighborhoods. *Economic Development Quarterly*, *26*(1), 50–72. http://dx.doi.org/10.1177/0891242411422902

Lövheim, D. (2014). *Science Education and Citizenship: Fairs, Clubs, and Talent Searches for American Youth, 1918–1958*, by Sevan G. Terzian. [review] *Science Education*, *98*(2), 368–70.

Low, S.M. (2000). *On the Plaza: The Politics of Public Space and Culture*. Austin, TX: University of Texas Press.

Lucas, M.J. (2014). The organizing practices of a community festival. *Journal of Organizational Ethnography*, *3*(2), 275–90. http://dx.doi.org/10.1108/JOE-01-2013-0001

Luchian, F. G. (2014). Audience research for the performing arts: Romanian Music Festival. *SEA-Practical Application of Science*, (2), 304–9.

Lucky, D., Turner, B., Hall, M., Lefaver, S., & de Werk, A. (2011). Blood pressure screenings through community nursing health fairs: Motivating individuals to seek health care follow-up. *Journal of Community Health Nursing*, *28*(3), 119–29. http://dx.doi.org/10.1080/07370016.2011.588589

Lukić Krstanović, M. (2011). Political folklore on festival market: Power of paradigm and power of stage. *Český lid: Etnologický časopis, 98*(3), 261–80.

Lyck, L., Long, P., & Griege, A.X. (2012). Introduction. In L. Lyck, P. Long & A.X. Grige (Eds.), *Tourism, Festivals and Cultural Events in Times of Crisis*, 1–6. Copenhagen: Fredenksberg bogtrykken.

Lytra, V. (2011). Negotiating language, culture and pupil agency in complementary school classrooms. *Linguistics and Education, 22*(1), 23–36. http://dx.doi.org/10.1016/j.linged.2010.11.007

Ma, E.A. (2014). *Hometown Chinatown: A History of Oakland's Chinese Community, 1852–1995*. New York: Routledge.

Ma, J., Song, W.G., Lo, S.M., & Fang, Z.M. (2013). New insights into turbulent pedestrian movement pattern in crowd-quakes. *Journal of Statistical Mechanics: Theory and Experiment*, (2). http://iopscience.iop.org/1742-5468/2013/02/P02028

MacLeod, G., & Johnstone, C. (2012). Stretching urban renaissance: Privatizing space, civilizing place, summoning "community." *International Journal of Urban and Regional Research, 36*(1), 1–28. http://dx.doi.org/10.1111/j.1468-2427.2011.01067.x

Mackellar, J.P. (2013a). Participant observation at events: Theory, practice and potential. *International Journal of Event and Festival Management, 4*(1), 56–65. http://dx.doi.org/10.1108/17582951311307511

Mackellar, J. (2013b). *Event Audiences and Expectations*. New York: Routledge.

Maclagan, D. (2010). *Outsider Art: From the Margins to the Marketplace*. London: Reaktion Books.

Maclean, K. (2010). Capitalizing on women's social capital? Women-targeted microfinance in Bolivia. *Development and Change, 41*(3), 495–515. http://dx.doi.org/10.1111/j.1467-7660.2010.01649.x

Madsen, J., Radel, C., & Endter-Wada, J. (2014). Justice and immigrant Latino recreation geography in Cache Valley, Utah. *Journal of Leisure Research, 46*(3), 291–312.

Madison, D.S. (2011). *Critical Ethnography: Method, Ethics, and Performance*. Thousand Oaks, CA: Sage.

Madyaningrum, M.E., & Sonn, C. (2011). Exploring the meaning of participation in a community art project: A case study on the Seeming project. *Journal of Community & Applied Psychology, 21*(4), 358–70. http://dx.doi.org/10.1002/casp.1079

Magazine, R. (2011). "We all put on the fiesta together": Interdependence and the production of active subjectivity through cargos in a highland Mexican village. *Journal of Latin American and Caribbean Anthropology, 16*(4), 296–314. http://dx.doi.org/10.1111/j.1935-4940.2011.01159.x

Mage, C., Goldstein, A.O., Colgan, S., Skinner, B., Kramer, K.D., Steiner, J., & Staples, A.H. (2010). Secondhand smoke policies at state and county fairs. *North Carolina Medical Journal, 71*(5), 409–12.

Magis, K. (2010). Community resilience: An indicator of social sustainability. *Society & Natural Resources: An International Journal, 23*(5), 401–16. http://dx.doi.org/10.1080/08941920903305674

Maguth, B.M., & Yamaguchi, M. (2010). Beyond the surface: A guide to substantive world fairs in the social studies. *Social Studies, 101*(2), 75–9. http://dx.doi.org/10.1080/00377990903284062

Main, D.S., & Velovis, A.J. (2010, Nov.). The effectiveness of boosting public health insurance enrollment through community events. The Colorado Trust. http://www.issuelab.org/resource/effectiveness_of_boosting_public_health_insurance_enrollment_through_community_events

Mahadi, Z., Hadi, A.S.A., & Sino, H. (2011). Public sustainable development values: A case study of Sepang, Malaysia. *Journal of Sustainable Development, 4*(2).

Main, K., & Sandoval, G.F. (2015). Placemaking in a translocal receiving community: The relevance of place identity and agency. *Urban Studies, 52*(1), 71–86.

Mair, J., & Laing, J. (2012). The greening of music festivals: Motivations, barriers and outcomes. Applying the Mair and Jago model. *Journal of Sustainable Tourism, 20*(5), 683–700.

Mair, J., & Whitford, M. (2013). An exploration of events research: Event topics, themes and emerging trends. *International Journal of Event and Festival Management, 4*(1), 6–30. http://dx.doi.org/10.1108/17582951311307485

Malek, A. (2011). Public performances of identity negotiation in the Iranian diaspora: The New York Persian Parade. *Comparative Studies of South Asia, Africa and the Middle East, 31*(2), 388–410. http://dx.doi.org/10.1215/1089201X-1264316

Mallach, A. (2010). *Bringing Buildings Back: From Abandoned Properties to Community Assets.* Montclair, NJ: National Housing Institute.

Mallee, H. & Pieke, F.N. (Eds.). (2014). *Internal and International Migration: Chinese Perspectives.* New York: Routledge.

Manthiou, A., Lee, S.A., Tang, L.R., & Chiang, L. (2014). The experience economy approach to festival marketing: Vivid memory and attendee loyalty. *Journal of Services Marketing, 28*(1), 22–35. http://dx.doi.org/10.1108/JSM-06-2012-0105

Mangia, G., Canonico, P., Toraldo, M.L., & Mercurio, R. (2011). Assessing the socio-economic impact of performing arts festivals: A new theoretical model. *Journal of US-China Public Administration, 8*(9), 1016–31.

Marais, M., & Saayman, M. (2011). Key success factors of managing the Robertson Wine Festival. *Acta Academica*, *43*(1), 146–66.

Marchi, R. (2009). *Day of the Dead in the USA: The Migration and Transformation of a Cultural Phenomenon*. Brunswick, NJ: Rutgers University Press.

Marcuse, P. (2014). The paradoxes of public space. *Journal of Architecture and Urbanism*, *38*(1), 102–6. http://dx.doi.org/10.3846/20297955.2014.891559

Margry, P.J., & Sanchez-Carretero, C. (2011). Rethinking memorialization: The concept of grassroots memorials. In P.J. Margry & C. Sanchez-Carretero (Eds.), *Grassroots Memorials: The Politics of Memorializing Traumatic Death*, 1–48. New York: Berghahn.

Markusen, A. (2007). *The Urban Core as Cultural Sticky Place*. Frankfurt: Time Space Places.

Markusen, A., & Gadwa, A. (2010). Arts and culture in urban and regional planning: A review and research agenda. *Journal of Planning Education and Research*, *29*(3), 379–91. http://dx.doi.org/10.1177/0739456X09354380

Markwell, K., & Tomsen, S. (2010). Safety and hostility at special events: Lessons from Australian gay and lesbian festivals. *Event Management*, *14*(3), 225–38. http://dx.doi.org/10.3727/152599510X12825895093678

Marler, B. (2011). Known cases of zoonotic pathogen outbreaks associated with state and county fairs, petting zoos, and community activities involving human-animal contact. *Fair and Petting Zoo Safety*. http://www.fair-safety.com/fair-outbreaks

Marquez, J.D. (2012). The Black Mohicans: Representations of everyday violence in postracial urban America. *American Quarterly*, *64*(3), 625–51. http://dx.doi.org/10.1353/aq.2012.0040

Marsden, M.T. (2010). The county fair as celebration and cultural text. *Journal of American Culture*, *33*(1), 24–9. http://dx.doi.org/10.1111/j.1542-734X.2010.00727.x

Marston, S.A. (1989). Public rituals and community power: St Patrick's day parades in Lowell, Massachusetts 1841–1874. *Political Geography Quarterly*, *8*(3), 255–69. http://dx.doi.org/10.1016/0260-9827(89)90041-4

Mason, J.D. (1996). Street fairs: Social space, social performance. *Theatre Journal*, *48*(3), 301–19. http://dx.doi.org/10.1353/tj.1996.0063

Masterman, G.R. (2004). A strategic approach for the use of sponsorship in the events industry: In search of a return on investment. In I. Yeoman, M. Robertson, J. Ali-Knight, S. Drummond, & U. McMahan-Beattle (Eds.), *Festival and Events Management: An International Arts and Culture Perspective*, 260–72. Oxford: Butterworth-Heinemann.

Maskell, P. (2014). Accessing remote knowledge—the roles of trade fairs, pipelines, crowdsourcing and listening posts. *Journal of Economic Geography*. http://joeg.oxfordjournals.org/content/early/2014/02/18/jeg.lbu002

Matheson, C.M., & Tinsley, R. (2014). The carnivalesque and event evolution: a study of the Beltane Fire Festival. *Leisure Studies*, 1–27. http://dx.doi.org/10.1080/02614367.2014.962591

Matthews, J. (2014). Voices from the heart: The use of digital storytelling in education. *Community Practitioner*, 87(1), 28–30.

Mattivi, R., Uijlings, J., De Natale, F.G., & Sebe, N. (2011). Exploitation of time constraints for (sub-)event recognition. In *Proceedings of the 2011 Joint ACM Workshop on Modeling and Representing Events*, 7–12. New York: ACM. http://dx.doi.org/10.1145/2072508.2072511

Mattivi, R., Uijlings, J., De Natale, F., & Sebe, N. (2012, June). Categorization of a collection of pictures into structured events. In *Proceedings of the 2nd ACM International Conference on Multimedia Retrieval*, 61. New York: ACM. http://dx.doi.org/10.1145/2324796.2324867

Mbaiwa, J.E. (2011). Cultural commodification and tourism: The Goo-Moremi Community, Central Botswana. *Tijdschrift voor Economische en Sociale Geografie*, 102(3), 290–301. http://dx.doi.org/10.1111/j.1467-9663.2011.00664.x

McCabe, S. (2006). The making of community identity through historic festive practice: The case of Ashbourne Royal Shrovetide Football. In D. Picard and M. Robinson (Eds.), *Festivals, Tourism and Social Change: Remaking Worlds*, 99–118. Clevedon: Channel View Publications.

McCarthy, B. (2013). The landscape of music festivals in Australia. In P. Tschmuck, P.L. Pearce, & S. Campbell (Eds.), *Music Business and the Experience Economy*, 119–34. Heidelberg: Springer. http://dx.doi.org/10.1007/978-3-642-27898-3_8

McClinchey, K.A. (2008). Urban ethnic festivals, neighborhoods, and the multiple realities of marketing place. *Journal of Travel & Tourism Marketing*, 25(3–4), 251–64. http://dx.doi.org/10.1080/10548400802508309

McComas, W.F. (2011). The science fair: A new look at an old tradition. *Science Teacher*, 78(8), 34–8.

McCrone, D., & McPherson, G. (Eds.). (2009). *National Days: Constructing and Mobilising National Identity*. Basingstoke: Palgrave Macmillan. http://dx.doi.org/10.1057/9780230251175

McDonald, J.F. (2013). What happened to and in Detroit? Social Science Research Network. http://papers.ssrn.com/sol3/papers.cfm?abstract_id=2344832

McDowall, S. (2011). The festival in my hometown: The relationships among performance quality, satisfaction, and behavioral intentions. *International Journal of Hospitality & Tourism Administration*, 12(4), 269–88. http://dx.doi.org/10.1080/15256480.2011.614528

McEachie, A. (2013). *Transcultural Transmission in Diasporic Festivals: Mas traditions in Brooklyn Labour Day and Notting Hill Carnival*. St Augustine, Trinidad & Tobago: University of the West Indies.

McGeough, D. (2013). Laboring for community, civic participation, and sanitation: The performance of Indian Toilet Festivals. *Text and Performance Quarterly*, *33*(4), 361–77. http://dx.doi.org/10.1080/10462937.2013.825925

McGregor, S., & Thompson-Fawcett, M. (2011). Tourism in a small town: Impacts on community solidarity. *International Journal of Sustainable Society*, *3*(2), 174–89. http://dx.doi.org/10.1504/IJSSOC.2011.039920

McGurgan, E., Robson, L., & Samenfink, W. (2012). The importance of site selection criteria for special events. Meeting Professionals International World Education Congress (WEC), Academic Forum 2012, 27–31 July, St. Louis, Missouri.

McKay, G., & Williams, R. (2010). Community Arts and Music, Community Media. In K. Howley (Ed.), *Understanding Community Media*, 41–52. London: Sage.

McPhedran, S. (2011). Disability and community life: Does regional living enhance social participation? *Journal of Disability Policy Studies*, *22*(1), 40–54. http://dx.doi.org/10.1177/1044207310394448

McQueen, C.P. (2010). Care of children at a large outdoor music festival in the United Kingdom. *Prehospital and Disaster Medicine*, *25*(3), 223–6.

Mcquire, S. (2010). Rethinking media events: Large screens, public space broadcasting and beyond. *New Media & Society*, *12*(4), 567–82. http://dx.doi.org/10.1177/1461444809342764

Mehmetoglu, M. (2001). Economic scale of community-run festivals: A case study. *Event Management*, *7*(2), 93–102. http://dx.doi.org/10.3727/152599501108751506

Meias, S., Pedersen, J.S., Kim, J.-H., Svejenova, S., & Mazza, C. (2011). Transforming film product identities: The status effects of European premier film festivals, 1996–2005. In. B. Moeran & J.S. Pedersen (Eds.), *Negotiating Values in the Creative Industries: Fairs, Festivals and Competitive Events*, 169–96. New York: Cambridge University Press.

Meites, E., & Brown, J.F. (2010). Ambulance need at mass gatherings. *Prehospital and Disaster Medicine*, *25*(6), 511–14. http://dx.doi.org/10.1017/S1049023X00008682

Melo, A.S. & Robeiro, M. do C.F. (2014). *Public Festivities in Portuguese Medieval Towns*. Bellaterra (Cerdanyola del Vallès): Institut d' Estudis Medievals (UAB-Spain).

Mendelsohn, J. (2013). *The Lower East Side Remembered and Revisited: History and Guide to a Legendary New York Neighborhood*. Updated and rev. New York: Columbia University Press.

Menjívar, C. (2006). Family reorganization in a context of legal uncertainty: Guatemalan and Salvadoran immigrants in the United States. *International Journal of Sociology of the Family*, *32*(2), 223–45.

Menon, J. (2013). Queer selfhoods in the shadow of neoliberal urbanism. *Journal of Historical Sociology*, 26(1), 100–19. http://dx.doi.org/10.1111/johs.12006

Mercier, C.G. (2007–08). Interpreting Brazilianness: Musical views of Brazil in Toronto. *MUSICultures: Journal of the Canadian Society for Traditional Music*, (34 and 35). https://journals.lib.unb.ca/index.php/MC/article/view/20258/23360

Merkel, U. (2013). The Grand Mass Gymnastics and Artistic Performance Arirang (2002–2012): North Korea's socialist-realist response to global sports spectacles. *International Journal of the History of Sport*, 30(11), 1247–58. http://dx.doi.org/10.1080/09523367.2013.793179

Merrilees, B., & Marles, K. (2011). Green business events: Profiling through a case study. *Event Management*, 15(4), 361–72. http://dx.doi.org/10.3727/152599511X13175676722609

Mielke, H.W., Gonzales, C.R., & Powell, E.T. (2012). *Potential Lead (Pb) Exposure from Carnival Beads and Parade Route Environments.* Report provided to Verdi Gras.

Milbourne, P. (2012). Everyday (in)justices and ordinary environmentalisms: Community gardening in disadvantaged urban neighbourhoods. *Local Environment*, 17(9), 943–57. http://dx.doi.org/10.1080/13549839.2011.607158

Miller, D. (2014). Migration, material culture and tragedy: Four moments in Caribbean migration. In J.J. Gieseking & W. Mangold (Eds.), *The People, Place, and Space Reader*, 100–3. New York: Routledge.

Milner, L.M., Jago, L.K., & Deery, M. (2004). Profiling the special event nonattendee: An initial investigation. *Event Management*, 8(3), 141–50. http://dx.doi.org/10.3727/1525995031436890

Miranda, M. (2010). *Power at the Roots: Gentrification, Community Gardens and the Puerto Ricans of the Lower East Side.* Lanham, MD: Lexington Books.

Mitchell, C. (2011). *Doing visual research.* London: Sage.

Mitchell, C., & de Lange, N. (2012). *Handbook of Participatory Video.* Walnut Creek, CA: AltaMira Press.

Moeran, B. (2011a). The book fair as a tournament of values. In B. Moeran & J.S. Pedersen (Eds.), *Negotiating Values in the Creative Industries: Fairs, Festivals and Competitive Events*, 119–44. New York: Cambridge University Press. http://dx.doi.org/10.1017/CBO9780511790393.006

Moeran, B. (2011b). Trade Fairs, Markets and Fields: Framing Imagined as Real Communities. *Historical Social Research/Historische Sozialforschung*, 36(3), 79–98.

Moeran, B. & Pedersen, J.S. (Eds.). (2011). *Negotiating Values in the Creative Industries: Fairs, Festivals and Competitive Events.* New York: Cambridge University Press. http://dx.doi.org/10.1017/CBO9780511790393

Molitor, F., Rossi, M., Branton, L., & Field, J. (2011). Increasing social capital and personal efficacy through small-scale community events. *Journal of Community Psychology*, *39*(6), 749–54. http://dx.doi.org/10.1002/jcop.20452

Molnar, D., & Kammerud, M. (1977). The problem analysis: The Delphi Technique. In N. Gilbert & H. Specht (Eds.), *Planning for Social Welfare*, 153–63. Englewood Cliffs, NJ: Prentice-Hall.

Monga, M. (2006). Measuring motivation to volunteer for special events. *Event Management*, *10*(1), 47–61. http://dx.doi.org/10.3727/152599506779364633

Montero, C.G. (2015). Tourism, cultural heritage and regional identities in the Isle of Spice. *Journal of Tourism and Cultural Change*, *13*(1), 1–21.

Montgomery, S.E., & Christie, E.M. (2011). A new take on New Year celebrations. *Social Studies and the Young Learner*, *24*(2), 14–18.

Mooney-Melvin, P. (2014). Engaging the neighborhood: The East Rogers Park Neighborhood History Project and the possibilities and challenges of community- based initiatives. *Journal of Urban History*, *40*(3), 462–78.

Moore, H.L.M. (2011). Laughing out loud: Art, culture, and fantasy. *Cardozo Law Review*, *33*, 2441.

Moore, K. (2014). What is next for participants of a Community Health Screening: Does screening increase medical home access? Paper presented at the 142nd APHA Annual Meeting and Exposition, 15–19 Nov., New Orleans.

Moore, R., Williamson, K., Sochor, M., & Brady, W.J. (2011). Large-event medicine – event characteristics impacting medical need. *American Journal of Emergency Medicine*, *29*(9), 1217–21. http://dx.doi.org/10.1016/j.ajem.2010.07.018

Moore, S.C., Flajšlik, M., Rosin, P.L., & Marshall, D. (2008). A particle model of crowd behavior: Exploring the relationship between alcohol, crowd dynamics and violence. *Aggression and Violent Behavior*, *13*(6), 413–22. http://dx.doi.org/10.1016/j.avb.2008.06.004

Morgan, K. (2010). Local and green, global and fair: The ethical foodscape and the politics of care. *Environment & Planning A*, *42*(8), 1852–67. http://dx.doi.org/10.1068/a42364

Moscardo, G. (2007). Analyzing the role of festivals and events in regional development. *Event Management*, *11*(1–2), 23–32. http://dx.doi.org/10.3727/152599508783943255

Muhammad, M., Wallerstein, N., Sussman, A.L., Avila, M., Belone, L., & Duran, B. (2014). Reflections on researcher identity and power: The impact of positionality on community based participatory research (CBPR) processes and outcomes. *Critical Sociology*. http://crs.sagepub.com/content/early/2014/06/11/0896920513516025.full.pdf

Mukwada, G., & Dhlamini, S. (2012). Challenges of event tourism in local economic development: The case of Bethlehem, South Africa. *Journal of Human Ecology*, *39*(1), 27–38.

Müller, M. (2014). *What Makes an Event a Mega-Event?* Definitions and Sizes. Social Science Research Network. http://papers.ssrn.com/sol3/papers.cfm?abstract_id=2462539

Munro, I., & Jordan, S. (2013). "Living space" at the Edinburgh Festival fringe: Spatial tactics and the politics of smooth space. *Human Relations*, *66*(11), 1497–525. http://dx.doi.org/10.1177/0018726713480411

Munson, C., & Chetkow-Yanoov, B.H. (2014). *Celebrating Diversity: Coexisting in a Multicultural Society*. New York: Routledge.

Murdie, R., & Teixeira, C. (2011). The impact of gentrification on ethnic neighbourhoods in Toronto: A case study of Little Portugal. *Urban Studies*, *48*(1), 61–83

Murgante, B., Tilio, L., Scorza, F., & Lanza, V. (2011). Crowd-cloud tourism, new approaches to territorial marketing. *Computational Science and Its Applications-ICCSA 2011*, *6783*, 265–76. Berlin: Springer. http://dx.doi.org/10.1007/978-3-642-21887-3_21

Murphy, D. (2014). Sport, culture and the media at the Festival Mondial des Arts Nègres de Dakar (2010): Sport and the democratisation of culture or sport as populism? *French Cultural Studies*, *25*(1), 10–22. http://dx.doi.org/10.1177/0957155813510691

Murphy, J.W. (2014). *Community-Based Interventions: Philosophy and Action*. New York: Springer.

Murray, D.A. (2014). Real queer: "Authentic" LGBT refugee claimants and homonationalism in the Canadian refugee system. *Anthropologica*, *56*(1), 21–32.

Murray, K., Liang, A., Barnack-Tavlaris, J., & Navarro, A.M. (2014). The reach and rationale for community health fairs. *Journal of Cancer Education*, *29*(1), 19–24. http://dx.doi.org/10.1007/s13187-013-0528-3

Murray, R. (2014, 13 Mar.). Corporations are rethinking sponsorships of Boston's St Patrick's Day Parade. GLAAD. http://www.glaad.org/blog/corporations-are-rethinking-sponsorship-bostons-exclusionary-st-patricks-day-parade

Myers, D., Budruk, M., & Andereck, K.L. (2010). Stakeholder involvement in destination level sustainable tourism indicator development: The case of a Southwestern U.S. mining town. In M. Burdruk & R. Phillips (Eds.), *Quality-of-Life Community Indicators for Parks, Recreation and Tourism Management*, 185–200.

Mykletun, R.J. (2011). Festival safety – Lessons learned from the Stavanger Food Festival (the Gladmatfestival). *Scandinavian Journal of Hospitality and Tourism*, *11*(3), 342–66. http://dx.doi.org/10.1080/15022250.2011.593363

Najafi, M., & Shariff, M.K.B.M. (2011). The concept of place and sense of place in architectural studies. *International Journal of Humanities and Social Science*, 6(3), 187–93.

Nakano, K. (2001). Flexibility of boundaries between "traditional" and "modern" groups: A case of separation of voluntary group in the urban festival, Kokura Gion Daikins. *Hagi International University Review* 4(3), 67–75.

Nam, C. (2012). Implications of community activism among urban minority young people for education for engaged and critical citizenship. *International Journal of Progressive Education*, 8(3), 62–76.

Nash, C.J., & Gorman-Murray, A. (2014). LGBT neighbourhoods and "new mobilities": Towards understanding transformations in sexual and gendered urban landscapes. *International Journal of Urban and Regional Research*, 38(3), 756–72. http://dx.doi.org/10.1111/1468-2427.12104

National Endowment for the Arts. (2010). *Live from Your Neighborhood: A National Study of Outdoor Arts Festivals*. Washington, DC: Author.

Németh, Z., Kuntsche, E., Urbán, R., Farkas, J., & Demetrovics, Z. (2011). Why do festival goers drink? Assessment of drinking motives using the DMQ-R SF in a recreational setting. *Drug and Alcohol Review*, 30(1), 40–6. http://dx.doi.org/10.1111/j.1465-3362.2010.00193.x

Newton, K.F. (2011). Arts activism: Praxis in social justice, critical discourse, and radical modes of engagement. *Art Therapy: Journal of the American Art Therapy Association*, 28(2), 50–6. http://dx.doi.org/10.1080/07421656.2011.578028

Newman, L., Dale, A., & Ling, C. (2011). Meeting on the edge: Urban spaces and the diffusion of the novel. *Spaces & Flows: An International Journal of Urban & ExtraUrban Studies*, 1(1), 1–14.

Newman, S.P. (1997). *Parades and the Politics of the Street: Festive Culture in the Early American Republic*. Philadelphia, PA: University of Pennsylvania Press. http://dx.doi.org/10.9783/9780812200478

Nicholson, H. (2012). The performance of memory: Drama, reminiscence and autobiography. *NJ (Drama Australia Journal)*, 36, 62–74.

Nilbe, K., Ahas, R., & Silm, S. (2014). Evaluating the travel distances of events visitors and regular visitors using mobile positioning data: The case of Estonia. *Journal of Urban Technology*, 21(2), 91–107. http://dx.doi.org/10.1080/10630732.2014.888218

Nishanth, T., Praseed, K.M., Rathnakaran, K., Satheesh Kumar, M.K., Ravi Krishna, R., & Valsaraj, K.T. (2012). Atmospheric pollution in a semi-urban, coastal region in India following festival seasons. *Atmospheric Environment*, 47(4), 295–306. http://dx.doi.org/10.1016/j.atmosenv.2011.10.062

Norton, B.L., McLeroy, K.R., Burdine, J.N., Felix, M.R.J., & Dorsey, A.M. (2002). Community capacity: Concept, theory, and methods. In R.J. DiClemente,

R.A. Crosby & M.C. Kegler (Eds.), *Emerging Theories in Health Promotion Practice and Research: Strategies for Improving Public Health*, 194–227. San Francisco, CA: Jossey-Bass.

Novelli, M. (2004). Wine tourism events, Apulia, Italy. In I. Yeoman, M. Robertson, J. Ali-Knight, S. Drummond, & U. McMahan-Beattle (Eds.), *Festival and Events Management: An International Arts and Culture Perspective*, 329–45. Oxford: Butterworth-Heinemann. http://dx.doi.org/10.1016/B978-0-7506-5872-0.50026-0

Novello, S., & Fernandez, P.M. (2014). The influence of event authenticity and quality attributes on behavioral intentions. *Journal of Hospitality & Tourism Research*. http://jht.sagepub.com/content/early/2014/01/05/1096348013515914.abstract

Nuere, C.O., & Ortuzar, A.M. (2013). Festive leisure: Significant changes in perception and organization of celebrations. *Tourism Research & Hospitality*, *2*(3).

Nunes, F. (2011). The Portuguese in Canada: Diasporic challenges and adjustment. [review] *Canadian Ethnic Studies*, *43*(1–2), 305–9. http://dx.doi.org/10.1353/ces.2011.0020

Nykiforuk, C.I.J., Vallianatos, H., & Nieuwendy, L.M. (2011). Photovoice as a method for revealing community perceptions of the built and social environment. *International Journal of Qualitative Methods*, *10*(2), 103–24.

O'Connor, C.D. (2013). Engaging young people? The experiences, challenges, and successes of Canadian Youth Advisory Councils. *Sociological Studies of Children and Youth*, *16*(1), 73–96. http://dx.doi.org/10.1108/S1537-4661(2013)0000016008

O'Donnell, C.R., & Tharp, R.G. (2012). Integrating cultural community psychology: Activity settings and the shared meanings of intersubjectivity. *American Journal of Community Psychology*, *49*(1), 22–30. http://dx.doi.org/10.1007/s10464-011-9434-1

O'Sullivan, D., & Jackson, M.J. (2002). Festival tourism: A contributor to sustainable local economic development. *Journal of Sustainable Tourism*, *10*(4), 325–42. http://dx.doi.org/10.1080/09669580208667171

O'Sullivan, D., Pickernell, D., & Senyard, J. (2009). Public sector evaluation of festivals and events. *Journal of Policy Research in Tourism, Leisure and Events*, *1*(1), 19–36. http://dx.doi.org/10.1080/19407960802703482

O'Toole, W. (2011). *Events Feasibility and Development: From Strategy to Operations*. Amsterdam: Elsevier.

Oakes, S., & Warnaby, G. (2011). Conceptualizing the management and consumption of live music in urban space. *Marketing Theory*, *11*(4), 405–18. http://dx.doi.org/10.1177/1470593111418798

Oberhagemann, D., Könnecke, R., & Schneider, V. (2014). Effect of social groups on crowd dynamics: Empirical findings and numerical simulations. In U. Weidmann, U. Kirsch, & M. Schreckenberg (Eds.), *Pedestrian and Evacuation Dynamics 2012*, 1251–8. Heidelberg: Springer. http://dx.doi.org/10.1007/978-3-319-02447-9_103

Okano, H., & Samson, D. (2010). Cultural urban branding and creative cities: A theoretical framework for promoting creativity in the public spaces. *Cities, 27*, S10–S15.

Okech, R.N. (2011). Promoting sustainable festival events tourism: A case study of Lamu Kenya. *Worldwide Hospitality and Tourism Themes, 3*(3), 193–202. http://dx.doi.org/10.1108/17554211111142158

Oldenburg, R. (1999). *The Great Good Place*. New York: Marlowe.

Oliveirinha, J., Pereira, F., & Alves, A. (2010). Acquiring semantic context for events from online resources. In *Proceedings of the 3rd International Workshop on Location and the Web*, 1. New York: ACM. http://dx.doi.org/10.1145/1899662.1899670

Olsen, C.S. (2013). Re-thinking festivals: A comparative study of the integration/marginalization of arts festivals in the urban regimes of Manchester, Copenhagen and Vienna. *International Journal of Cultural Policy, 19*(4), 481–500. http://dx.doi.org/10.1080/10286632.2012.661420

Onyx, J., & Leonard, R. (2010). The conversion of social capital into community development: An intervention in Australia's outback. *International Journal of Urban and Regional Research, 34*(2), 381–97. http://dx.doi.org/10.1111/j.1468-2427.2009.00897.x

Ore, T.E. (2011). Something from nothing: Women, space, and resistance. Gender & Society, 25(6), 689–95. http://dx.doi.org/10.1177/0891243211425176

Orsi, R.A. (2010). *The Madonna of 115th Street: Faith and Community in Italian Harlem, 1880–1950*. New Haven, CT: Yale University Press.

Oswald, J. (2005). Job Fairs in America's State Prisons: Summary of Findings on Research. *Journal of Correctional Education, 56*(2), 174–85.

Otero, L.R. (2010). *La Calle: Spatial Conflicts and Urban Renewal in a Southwest City*. Tucson, AZ: University of Arizona Press.

Ottawa Chinatown. (2014). Chinatown Lunar New Year Dance Parade. http://www.ottawachinatown.ca

Packer, J., & Ballantyne, J. (2011). The impact of music festival attendance on young people's psychological and social well-being. *Psychology of Music, 39*(2), 164–81. http://dx.doi.org/10.1177/0305735610372611

Packman, J. (2012). The Carnavalização of São João: Forrós, sambas and festive Interventions during Bahia, Brazil's festas juninas. *Ethnomusicology Forum, 21*(3), 327–53.

Padilla, R., Bull, S., Raghunath, S.G., Fernald, D., Havranek, E.P., & Steiner, J.F. (2010). Designing a cardiovascular disease prevention web-site for Latinos: Qualitative community feedback. *Health Promotion Practice*, *11*(1), 140–7. http://dx.doi.org/10.1177/1524839907311051

Palma, M., Palma, L., & Aguado, L. (2013). Determinants of cultural and popular celebration attendance: The case study of Seville Spring Fiestas. *Journal of Cultural Economics*, *37*(1), 87–107. http://dx.doi.org/10.1007/s10824-012-9167-5

Pamuk, A. (2004). Geography of immigrant clusters in global cities: A case study of San Francisco, 2000. *International Journal of Urban and Regional Research*, *28*(2), 287–307. http://dx.doi.org/10.1111/j.0309-1317.2004.00520.x

Pan, L. (2014). Who is occupying wall and street: Graffiti and urban spatial politics in contemporary China. *Continuum (Perth)*, *28*(1), 136–53. http://dx.doi.org/10.1080/10304312.2013.854867

Park, J., Lee, G., & Park, M. (2011). Service quality dimensions perceived by film festival visitors. *Event Management*, *15*(1), 49–61. http://dx.doi.org/10.3727/152599511X12990855575141

Parker, G., & Doak, J. (2012). *Key Concepts in Planning*. Los Angeles: Sage. http://dx.doi.org/10.4135/9781473914629

Patel, T.G. (2014). "We'll go grafting, yea?": Crime as a response to urban unrest. *Criminology & Criminal Justice*, *14*(2), 179–95.

Patterson, A., Cromby, J., Brown, S.D., Gross, H., & Locke, A. (2011). "It all boils down to respect doesn't it?": Enacting a sense of community in a deprived inner-city area. *Journal of Community & Applied Social Psychology*, *21*(4), 342–57. http://dx.doi.org/10.1002/casp.1078

Pauwels, L. (2011). An integrated conceptual framework for visual social research. In *The SAGE Handbook of Visual Research Methods*, 3. Thousand Oaks, CA: Sage.

Pearce, S.C., & Cooper, A. (2014). LGBT movements in Southeast Europe: Violence, justice, and international intersections. In *Handbook of LGBT Communities, Crime, and Justice*, 311–38. New York: Springer.

Pearlman, L.A. (2013). Restoring self in community: Collective approaches to psychological trauma after genocide. *Journal of Social Issues*, *69*(1), 111–24. http://dx.doi.org/10.1111/josi.12006

Pehrson, S., Stevenson, C., Muldoon, O.T., & Reicher, S. (2014). Is everyone Irish on St Patrick's Day? Divergent expectations and experiences of collective self- objectification at a multicultural parade. *British Journal of Social Psychology*, *53*(2), 249–64.

Peng, W., & Tang, L. (2010). Health content in Chinese newspapers. *Journal of Health Communication*, *15*(7), 695–711. http://dx.doi.org/10.1080/10810730.2010.514028

Pennay, A., & Room, R. (2012). Prohibiting public drinking in urban public spaces: A review of the evidence. *Drugs Education Prevention & Policy, 19*(2), 91–101. http://dx.doi.org/10.3109/09687637.2011.640719

Pentecost, R., Spence, M., & Kale, S. (2011). Events gone bad: Ramifications and theoretical reasoning. *International Journal of Sport and Society, 2*(2), 29–40.

Pepe, P.E., & Nichols, S. (2013). Event medicine: An evolving subspecialty of emergency medicine. In J.-L. Vincent (Ed.), *Annual Update in Intensive Care and Emergency Medicine 2013*, 37–47. Berlin: Springer. http://dx.doi.org/10.1007/978-3-642-35109-9_3

Percy-Smith, B., & Carney, C. (2011). Using art installations as action research to engage children and communities in evaluating and redesigning city centre spaces. *Educational Action Research, 19*(1), 23–39. http://dx.doi.org/10.1080/09650792.2011.547406

Pereira, F.C., Rodrigues, F., & Ben-Akiva, M. (2015). Using data from the web to predict public transport arrivals under special events scenarios. *Journal of Intelligent Transportation Systems: Technology, Planning, and Operations, 19*(3), 273–88.

Peter, S., Anandkumar, V. & Peter, S. (2013). Role of shopping festivals in destination branding: A tale of two shopping festivals in the United Arab Emeries. *Anatolia: An International Journal of Tourism and Hospitality Research, 24*(2), 264–67.

Peters, A. (2014). Festivals as a celebration of place: A case study of the Coranderrk Aboriginal festival. *CAUTHE 2014: Tourism and Hospitality in the Contemporary World: Trends, Changes and Complexity*, 481–95. Brisbane: School of Tourism, The University of Queensland.

Peters, K., & de Haan, H. (2011). Everyday spaces of inter-ethnic interaction: The meaning of urban public spaces in the Netherlands. *Leisure/Loisir, 35*(2), 169–90.

Phan, M.B., & Luk, C.M. (2008). "I don't say I have a business in Chinatown": Chinese sub-ethnic relations in Toronto's Chinatown West. *Ethnic and Racial Studies, 31*(2), 294–326. http://dx.doi.org/10.1080/01419870701342379

Phi, G., Dredge, D., & Whitford, M. (2014). Understanding conflicting perspectives in event planning and management using Q method. *Tourism Management, 40*, 406–15. http://dx.doi.org/10.1016/j.tourman.2013.07.012

Phipps, P., & Slater, L. (2010). *Indigenous cultural festivals: Evaluating impact on community health and wellbeing*. Victoria: Globalism Research Centre, RMIT University.

Pine, J.B., & Gilmore, J.H. (2007). *Authenticity: What Consumers Really Want?* Cambridge, MA: Harvard Business School.

Pipes, M.-L. (2013). Evidence of public celebrations and feasting: Politics and agency in late eighteenth-early nineteenth century New York. In M.F. Janowitz & D. Dallal (Eds.), *Tales of Gotham, Historical Archaeology, Ethnohistory and Microhistory of New York*, 265–83. New York: Springer.

Platt, L. (2012). "Parks are dangerous and the sidewalk is closer": Children's use of neighborhood space in Milwaukee, Wisconsin. *Children, Youth and Environments*, 22(2), 194–213.

Pochettino, M.L., Puentes, J.P., Buct Costantino, F., Arenas, P.M., Ulibarri, E.A., & Hurrell, J.A. (2012). Functional foods and nutraceuticals in a market of Bolivian immigrants in Buenos Aires (Argentina). *Evidence-Based Complementary and Alternative Medicine*, art. ID 320193.

Poirier, C. (2011). Proactive cities and cultural diversity: Policy issues and dynamics. *Our Diverse Cities*, 7(Spring), 24–9.

Poling, J., & Thalheimer, E. (2011). Community noise agreements, monitoring, and control for concerts on Boston's Rose Kennedy Greenway. Presented at NOISE-CON 2011, 25–7 July, Portland, Oregon.

Poortinga, W. (2012). Community resilience and health: The role of bonding, bridging, and linking aspects of social capital. *Health & Place*, 18(2), 286–95. http://dx.doi.org/10.1016/j.healthplace.2011.09.017

Porter, L. & Shaw, K. (Eds.). (2013). *Whose Urban Renaissance?: An International Comparison of Urban Regeneration Strategies*. New York: Routledge.

Portes, A. (2012). Tensions that make a difference: Institutions, interests, and the immigrant drive. *Sociological Forum*, 27(3), 563–78. http://dx.doi.org/10.1111/j.1573-7861.2012.01335.x

Powell, E., Stukes, F., Barnes, T., & Lipford, H.R. (2011). Snag'em: Creating community connections through games. In *Privacy, Security, Risk and Trust (PASSAT), 2011 IEEE Third International Conference on Social Computing (SocialCom)*, 591–4. New York: IEEE.

Premdas, R. (2004). Diaspora and its discontents: A Caribbean fragment in Toronto in quest of cultural recognition and political empowerment. *Ethnic and Racial Studies*, 27(4), 544–64. http://dx.doi.org/10.1080/01491987042000216708

Prentice, R., & Andersen, V. (2003). Festival as creative destination. *Annals of Tourism Research*, 30(1), 7–30. http://dx.doi.org/10.1016/S0160-7383(02)00034-8

Price, P.L. (2010). Cultural geography and the stories we tell ourselves. *Cultural Geographies*, 17(2), 203–10. http://dx.doi.org/10.1177/1474474010363851

Prilleltensky, I. (2012). Wellness as fairness. *American Journal of Community Psychology*, 49(1–2), 1–21. http://dx.doi.org/10.1007/s10464-011-9448-8

Pretorius, M., Gwynne, S., & Galea, E.R. (2015). Large crowd modelling: An analysis of the Duisburg Love Parade disaster. *Fire and Materials, 39*(4), 301–22.

Procter, D.E. (2004). Building community through communication: The case for civic communion. *Community Development, 35*(2), 53–72. http://dx.doi.org/10.1080/15575330409490132

Procter, D.E. (2006). *Civic Communion: The Rhetoric of Community Building.* Lantham, MD: Rowman & Littlefield.

Quan-Haase, A., & Martin, K. (2013). Digital curation and the networked audience of urban events Expanding La Fiesta de Santo Tomás from the physical to the virtual environment. *International Communication Gazette, 75*(5–6), 521–537. http://dx.doi.org/10.1177/1748048513491910

Qadeer, M.A., & Agrawal, S.K. (2011). The practice of multicultural planning in American and Canadian cities. *Canadian Journal of Urban Research, 20*(1) (Suppl), 132–56.

Quick, K.S., & Feldman, M.S. (2011). Distinguishing participation and inclusion. *Journal of Planning Education and Research, 31*(3), 272–90. http://dx.doi.org/10.1177/0739456X11410979

Quinn, B. (2005). Arts festivals and the city. *Urban Studies (Edinburgh, Scotland), 42*(5), 927–43. http://dx.doi.org/10.1080/00420980500107250

Quinn, B. (2006). Problematising "festival tourism": Arts festivals and sustainable development in Ireland. *Journal of Sustainable Tourism, 14*(3), 288–306. http://dx.doi.org/10.1080/09669580608669060

Quinn, N., Shulman, A., Knifton, L., & Byrne, P. (2011). The impact of a national mental health arts and film festival on stigma and recovery. *Acta Psychiatrica Scandinavica, 123*(1), 71–81. http://dx.doi.org/10.1111/j.1600-0447.2010.01573.x

Racine, G., Truchon, K., & Hage, M. (2008). And we are still walking ... When a protest walk becomes a step towards research on the move. *FQS: Forum Qualitative Sozialforschung/Forum: Qualitative Social Research, 9*(2).

Raeburn, J., Akerman, M., Chuengsatiansup, K., Mejia, F., & Oladepo, O. (2006). Community capacity building and health promotion in a globalized world. *Health Promotion International, 21*(Suppl 1), 84–90. http://dx.doi.org/10.1093/heapro/dal055

Raj, R., Walters, P., & Rashid, T. (2013). *Events Management: Principles and Practice.* 2nd ed. Thousand Oaks, CA: Sage.

Ramos, N. (2014, 8 Sept.). City used high-tech tracking software. *Boston Globe*, B1–B2.

Rapkiewicz, A.V., Shuman, M.J., & Hutchins, K.D. (2014). Fatal wounds sustained from "falling bullets": Maintaining a high index of suspicion in a forensic setting. *Journal of Forensic Sciences, 59*(1), 268–70. http://dx.doi.org/10.1111/1556-4029.12258

Rast, R.W. (2007). The cultural politics of tourism in San Francisco's Chinatown, 1882–1917. *Pacific Historical Review, 76*(1), 29–60. http://dx.doi.org/10.1525/phr.2007.76.1.29

Ratcliff, J.J., Miller, A.K., & Krolikowski, A.M. (2013). Why Pride displays elicit support from majority group members: The mediational role of perceived deservingness. *Group Processes & Intergroup Relations, 16*(4), 462–75. http://dx.doi.org/10.1177/1368430212453630

Rath, J. (2007). *The Transformation of Ethnic Neighborhoods into Places of Leisure and Consumption.* San Diego: Center for Comparative Immigration Studies, University of California.

Ratts, M.J. (2011). Multiculturalism and social justice: Two sides of the same coin. *Journal of Multicultural Counseling and Development, 39*(1), 24–37. http://dx.doi.org/10.1002/j.2161-1912.2011.tb00137.x

Ray, B. (2004). A diversity paradox: Montréal's gay village. *Our Diverse Cities, 1*(1), 72–5.

Raymond, C.M., Brown, G., & Weber, D. (2010). The measurement of place attachment: Personal, community, and environmental connections. *Journal of Environmental Psychology, 30*(4), 422–34. http://dx.doi.org/10.1016/j.jenvp.2010.08.002

Recio Mir, Á., & Cinelli, N. (2013). Art, celebration and fire: The Cathedral of Seville and the fireworks in the second half of the 17th century. *Ricerche di storia dell'arte, 36*(1), 87–97.

Reid, S. (2004). The social consequences of rural events: The Inglewood Olive Festival. http://www98.griffith.edu.au/dspace/bitstream/handle/10072/47967/55931_1.pdf?sequence=1

Reid, S. (2011). Event stakeholder management: developing sustainable rural event practices. *International Journal of Event and Festival Management, 2*(1), 20–36. http://dx.doi.org/10.1108/17582951111116597

Reitz, J.G., & Lum, J.M. (2006). Immigration and diversity in a changing Canadian city: Social bases of intergroup relations in Toronto. In E. Fong (Ed.), *Inside the Mosaic,* 15–50. Toronto: University of Toronto Press.

Reverte, F.G., & Izard, O.M. (2011). The role of social and intangible factors in cultural event planning in Catalonia. *International Journal of Event and Festival Management, 2*(1), 37–53. http://dx.doi.org/10.1108/17582951111116605

Revet, S. (2011). Remembering La Tragedia: Commemorations of the 1999 floods in Venezuela. In P.J. Margry & C. Sanchez-Carretero (Eds.), *Grassroots Memorials: The Politics of Memorializing Traumatic Death,* 208–25. New York: Berghahn.

Riaño-Alcalá, P., & Goldring, L. (2014). Unpacking refugee community transnational organizing: The challenges and diverse experiences of Colombians in Canada. *Refugee Survey Quarterly, 33*(2), 84–111.

Richardson, T.A. (2011). At the garden gate: Community building through food – revisiting the critique of "food, folk and fun" in multicultural education. *Urban Review, 43*(1), 107–23. http://dx.doi.org/10.1007/s11256-009-0146-x

Rickly-Boyd, J.M. (2012). Existential authenticity: Place matters. *Tourism Geographies: An International Journal of Tourism Space, Place and Environment, 15*(4), 680–6.

Ridge, J.T., & Bushnell, L. (2011). *Celebrating 250 years of the New York City St. Patrick's Day Parade.* Hamden, CT: Quinnipiac University Press.

Rihova, I. (2013). *Customer-to-Customer Co-creation of Value in the Context of Festivals.* Doctoral dissertation, Bournemouth University.

Robertson, M., Rogers, P., & Leask, A. (2009). Progessing socio-cultural impact evaluation for festivals. *Journal of Policy Research in Tourism, Leisure and Events, 1*(2), 151–69. http://dx.doi.org/10.1080/19407960902992233

Robbins, C.R., & Robbins, M.W. (2014). Engaging the contested memory of the public square: Community collaboration, archaeology, and oral history at Corpus Christi's Artesian Park. *Public Historian, 36*(2), 26–50. http://dx.doi.org/10.1525/tph.2014.36.2.26

Roberts, S. (2011, 21 Feb.). New York's Little Italy, littler by the year. *New York Times*, 21.

Rocha-Trindade, M.B. (2009). The Portuguese diaspora. In C. Teixeira and V.M.P. Da Rosa (Eds.), *The Portuguese in Canada: Diasporic challenges and adjustment*, 18–42. Toronto: University of Toronto Press.

Roemer, M.A. (2007). Ritual participation and social support in a major Japanese festival. *Journal for the Scientific Study of Religion, 46*(2), 185–200. http://dx.doi.org/10.1111/j.1468-5906.2007.00350.x

Rogers, P., & Anastasiadou, C. (2011). Community involvement in festivals: Exploring ways of increasing local participation. *Event Management, 15*(4), 387–99. http://dx.doi.org/10.3727/152599511X13175676722681

Rogers, P., & Barron, P. (2010). Finding new ways of evaluating the socio-cultural impacts of festivals and events. *CAUTHE 2010: Tourism and hospitality: Challenge the limits*, 1256–1260.

Rong-Da Liang, A., Chen, S.C., Tung, W., & Hu, C.C. (2013). The influence of food expenditure on tourist response to festival tourism: Expenditure perspective. *International Journal of Hospitality & Tourism Administration, 14*(4), 377–97. http://dx.doi.org/10.1080/15256480.2013.838088

Rosales, R. (2013). Survival, economic mobility and community among Los Angeles fruit vendors. *Journal of Ethnic and Migration Studies, 39*(5), 697–717. http://dx.doi.org/10.1080/1369183X.2013.756659

Rosendahl, T.J. (2012). *Music and Queer Culture: Negotiating Marginality through Musical Discourse at Pride Toronto*. Doctoral dissertation, Florida State University. http://diginole.lib.fsu.edu/etd/5427

Rosenstein, C. (2011). Cultural development and city neighborhoods. *City, Culture and Society*, 2(1), 9–15.

Rosental, C. (2013). Toward a sociology of public demonstrations. *Sociological Theory*, 31(4), 343–65.

Ross, T. (2013). "Telling the Brown stories": An examination of identity in the ethnic media of multigenerational immigrant communities. *Journal of Ethnic and Migration Studies*, 40(8), 1314–29.

Ross, L. F., Loup, A., Nelson, R. M., Botkin, J. R., Kost, R., Smith Jr, G. R., & Gehlert, S. (2010). The challenges of collaboration for academic and community partners in a research partnership: Points to consider. *JERHRE: Journal of Empirical Research on Human Research Ethics*, 5(1), 19–31.

Rothblum, E.D. (2014). Mars to Venus or earth to earth? How do families of origin fit into GLBTQ lives? *Journal of GLBT Family Studies*, 10(1–2), 231–41. http://dx.doi.org/10.1080/1550428X.2014.857235

Rothstein, E. (2014, 3 Oct.). Great job on the railroad, now go back to China. *New York Times*, C21, C27.

Roy, C. (2013). Organizers of cultural events: Creating community and telling stories of resistance and change. In C. Kawalilak and J. Groen (Eds.), *CASAE/ACEEA Conference Proceedings University of Victoria British Columbia, June 3–5, 2013*, 521–7. Victoria: University of Victoria.

Ruling, C.-C. (2011). Event institutionalization and maintenance: The Annecy Animation Festival 1960–2010. In B. Moeran & J.S. Pedersen (Eds.), *Negotiating Values in the Creative Industries: Fairs, Festivals and Competitive Events*, 197– 223. New York: Cambridge University Press. http://dx.doi.org/10.1017/CBO9780511790393.009

Ruprecht, T. (2010). *Toronto's Many Faces*. Toronto: Dundurn.

Rushford, M.-A. (2013). The where the heart is ... community festival: Celebrating community connections. *Parity*, 26(1), 44–maps5.

Ryan, D. (2000). *Orange Parades: The Politics of Ritual, Tradition and Control*. London: Pluto.

Ryan, A.W., & Wollan, G. (2013). Festivals, landscapes, and aesthetic engagement: A phenomenological approach to four Norwegian festivals. *Norwegian Journal of Geography*, 67(2), 99–112.

Ryan, J.S. (2006). The range – and purposes – of Australian public festivals that are functioning at present. *Folklore: Electronic Journal of Folklore*, (34), 7–30.

Saayman, M., Kruger, M., & Erasmus, J. (2012). Finding the key to success: A visitors' perspective at a National Arts Festival. *Acta Commercii: Independent Research Journal in the Management Sciences, 12*(1), 150–72.

Saayman, M., & Saayman, A. (2006). Does the location of arts festivals matter for the economic impact? *Papers in Regional Science, 85*(4), 569–84. http://dx.doi.org/10.1111/j.1435-5957.2006.00094.x

Sadd, D. (2012). *Mega-events, Community Stakeholders and Legacy: London 2012.* Doctoral dissertation, Bournemouth University.

Sakakeeny, M. (2010). "Under the bridge": An orientation to soundscapes in New Orleans. *Ethnomusicology, 54*(1), 1–27. http://dx.doi.org/10.5406/ethnomusicology.54.1.0001

Salazar, N.B. (2012). Community-based cultural tourism: Issues, threats, and opportunities. *Journal of Sustainable Tourism, 20*(1), 9–22. http://dx.doi.org/10.1080/09669582.2011.596279

Salem, G., Jones, E., & Morgan, N. (2004). An overview of arts management. In I. Yeoman, M. Robertson, J. Ali-Knight, S. Drummond, & U. McMahan-Beattle (Eds.), *Festival and events management: An International arts and culture perspective,* 14–31. Oxford: Butterworth-Heinemann. http://dx.doi.org/10.1016/B978-0-7506-5872-0.50007-7

Salzbrunn, M. (2014). How diverse is Cologne carnival? How immigrants appropriate popular art spaces. *Identities: Global Studies in Culture and Power, 21*(1), 92–106. http://dx.doi.org/10.1080/1070289X.2013.841581

Samadi, Z., Yunus, R.M., & Omar, D. (2012). Evaluating revitalizing toolkit towards a quality heritage establishment. *Procedia: Social and Behavioral Sciences, 35,* 637–44. http://dx.doi.org/10.1016/j.sbspro.2012.02.131

Sandoval, G.F., & Maldonado, M.M. (2012). Latino urbanism revisited: Placemaking in new gateways and the urban-rural interface. *Journal of Urbanism: International Research on Placemaking and Urban Sustainability, 5*(2–3), 193–218.

Sanjek, R. (2014). *Ethnography in Today's World: Color Full before Color Blind.* Philadelphia, PA: University of Pennsylvania Press.

Santino, J. (2011). The carnivalesque and the ritualesque. *Journal of American Folklore, 124*(491), 61–73. http://dx.doi.org/10.5406/jamerfolk.124.491.0061

Sarkissian, W. (2010). The Beginning of Something": Using Video as a Tool in Community Engagement. *Multimedia Explorations in Urban Policy and Planning: Urban and Landscape Perspective, 7,* 151–65. http://dx.doi.org/10.1007/978-90-481-3209-6_8

Sarmiento, C.S., & Beard, V.A. (2013). Traversing the border: Community-based planning and transnational migrants. *Journal of Planning Education and Research, 33*(3), 336–47. http://dx.doi.org/10.1177/0739456X13499934

Sassatelli, M. (2011). Urban festivals and the cultural public space. In L. Giorgi, M. Sassatelli, & G. Delanty (Eds.), *Festivals and the Cultural Public Sphere*, 12–28. London: Routledge.

Sassen, S. (2013). Informal knowledge and its enablement: The role of the new technologies. In F.D.-N. Rubio & P. Baert (Eds.), *The Politics of Knowledge*, 96–117. New York: Routledge.

Scambary, J. (2013). Conflict and resilience in an urban squatter settlement in Dili, East Timor. *Urban Studies*, 50(10), 1935–50.

Scerri, A., & James, P. (2010). Communities of citizens and "indicators" of sustainability. *Community Development Journal: An International Forum*, 45(2), 219–36. http://dx.doi.org/10.1093/cdj/bsp013

Schiller, N.G. (2012a). Unraveling the migration and development web: Research and policy implications. *International Migration*, 50(3), 92–7. http://dx.doi.org/10.1111/j.1468-2435.2012.00757.x

Schiller, N.G. (2012b). A comparative relative perspective on the relationships between migrants and cities. *Urban Geography*, 33(6), 879–903. http://dx.doi.org/10.2747/0272-3638.33.6.879

Schippers, H., & Bartleet, B.L. (2013). The nine domains of community music: Exploring the crossroads of formal and informal music education. *International Journal of Music Education*, 31(4), 454–71. http://dx.doi.org/10.1177/0255761413502441

Schmallegger, D., & Carson, D. (2010). Whose tourism city is it? The role of government in tourism in Darwin, Northern Territory. *Tourism and Hospitality Planning & Development*, 7(2), 111–29. http://dx.doi.org/10.1080/14790531003737144

Schnoor, R.F. (2006). Being gay and Jewish: Negotiating intersecting identities. *Sociology of Religion*, 67(1), 43–60. http://dx.doi.org/10.1093/socrel/67.1.43

Schulenkorf, N. (2012). Sustainable community development through sport and events: A conceptual framework for Sport-for-Development projects. *Sport Management Review*, 15(1), 1–12. http://dx.doi.org/10.1016/j.smr.2011.06.001

Schulenkorf, N., & Edwards, D. (2012). Maximizing positive social impacts: Strategies for sustaining and leveraging the benefits of inter-community sport events in divided societies. *Journal of Sport Management*, 26(5), 379–90.

Schulenkorf, N., Thomson, A., & Schlenker, K. (2011). Intercommunity sport events: Vehicles and catalysts for social capital in divided societies. *Event Management*, 15(2), 105–19. http://dx.doi.org/10.3727/152599511X13082349958316

Schulte-Römer, N. (2013). Fair framings: Arts and culture festivals as sites for technical innovation. *Mind & Society*, 12(1), 151–65. http://dx.doi.org/10.1007/s11299-013-0114-8

Scully, M. (2012). Whose day is it anyway? St Patrick's Day as a contested performance of national and diasporic Irishness. *Studies in Ethnicity and Nationalism, 12*(1), 118–35. http://dx.doi.org/10.1111/j.1754-9469.2011.01149.x

Schwarz, A. (2012). How publics use social media to respond to blame games in crisis communication: The Love Parade tragedy in Duisburg 2010. *Public Relations Review, 38*(3), 430–7. http://dx.doi.org/10.1016/j.pubrev.2012.01.009

Schwarz, E.C., & Tait, R. (2007). Recreation, arts, events and festivals: Their contribution to a sense of community in the Colac-Otway Shire of country Victoria. *Rural Society, 17*(2), 125–38. http://dx.doi.org/10.5172/rsj.351.17.2.125

Seo, D.C. (2011). Lessons learned from a Black and minority health fair's 15-month follow-up counseling. *Journal of the National Medical Association, 103*(9–10), 897–906.

Sengupta, A.W. (2012). *Immigration, Ethnicity and National Identity in Chicago, Illinois.* Dissertation, MIT Urban Studies Department. http://hdl.handle.net/1721.1/70414.

Sepe, M. (2013). *Planning and Place in the City: Mapping Place Identity.* New York: Routledge.

Seymour, M., Wolch, J., Reynolds, K.D., & Bradbury, H. (2010). Resident perceptions of urban alleys and alley gardening. *Applied Geography, 30*(3), 380–93. http://dx.doi.org/10.1016/j.apgeog.2009.11.002

SF Gate (2014, 2 July). *San Francisco: Chinatown.* http://www.sfgate.com/neighborhoods/sf/chinatown/

Shah, S. (2014). Corporate social responsibility: A way of life at the Tata Group. *Journal of Human Values, 20*(1), 59–74. http://dx.doi.org/10.1177/0971685813515591

Sharpley, P.R., & Stone, P. (2011). Socio-cultural impacts of events: meanings, authorised transgression and social capital. In S. Page & J. Connell (Eds.), *The Routledge Handbook of Events,* 347–61. London: Routledge.

Shaw, A., & Ardener, S. (Eds.) (2005). *Changing Sex and Bending Gender.* New York: Berghahn.

Shin, H.B. (2012). Unequal cities of spectacle and mega-events in China. *City: Analysis of Urban Trends, Culture, Theory, Policy Action, 16*(6), 728–44.

Shragge, E. (2013). *Activism and Social Change: Lessons from Community Organizing.* 2nd ed. Toronto: University of Toronto Press.

Shutika, D.L. (2008). The ambivalent welcome: Cinco de Mayo and the symbolic expression of local identity and ethnic relations. In D.S. Massey (Ed.), *New Faces in New Places: The Changing Geography of American Immigration,* 274–307. New York: Russell Sage Foundation.

Siemiatycki, M., & Isin, E.F. (1997). Immigration, diversity and urban citizenship in Toronto. *Canadian Journal of Regional Science, 20*(1), 73–102.

Sievert, T. (2006). *Chinese New Year: Festival of New Beginnings*. Mankato, MN: Capstone.

Silverman, E. (2010). A reason to celebrate: Recognizing accomplishments in good times and bad. *Bottom Line: Managing Library Finances, 23*(1), 44–6. http://dx.doi.org/10.1108/08880451011049704

Silvers, J.R., & Goldblatt, J. (2012). *Professional Event Coordination*. New York: Wiley.

Simao, J.N., & Partidario, M.R. (2012). How does tourism planning contribute to sustainable development? *Sustainable Development, 20*(6), 372–85. http://dx.doi.org/10.1002/sd.495

Simon, N. (2010). *The Participatory Museum*. http://www.participatorymuseum.org/

Sites, W., Chaskin, R.J., & Parks, V. (2007). Reframing community for the 21st century: Multiple traditions, multiple challenges. *Journal of Urban Affairs, 29*(5), 519–41. http://dx.doi.org/10.1111/j.1467-9906.2007.00363.x

Sjollema, S.D., & Hanley, J. (2014). When words arrive: A qualitative study of poetry as a community development tool. *Community Development Journal: An International Forum, 49*(1), 54–68. http://dx.doi.org/10.1093/cdj/bst001

Skerratt, S., & Hall, C. (2011). Community ownership of physical assets: Challenges, complexities and implications. *Local Economy, 26*(3), 170–81. http://dx.doi.org/10.1177/0269094211401491

Skot-Hansen, D., Rasmussen, C.H., & Jochumsen, H. (2013). The role of public libraries in culture-led urban regeneration. *New Library World, 114*(1/2), 7–19. http://dx.doi.org/10.1108/03074801311291929

Skov, L., & Meier, J. (2011). Configuring sustainability at fashion week. In B. Moeran & J.S. Pedersen (Eds.), *Negotiating Values in the Creative Industries: Fairs, Festivals and Competitive Events*, 270–93. New York: Cambridge University Press. http://dx.doi.org/10.1017/CBO9780511790393.012

Slabbert, E., & Saayman, M. (2011). The influence of culture on community perceptions: The case of two South African arts festivals. *Event Management, 15*(2), 197–211. http://dx.doi.org/10.3727/152599511X13082349958352

Slater, L. (2010). "Calling our Spirits Home": Indigenous cultural festivals and the making of a good life. *Cultural Studies Review, 19*(1), 43–54.

Slater, L. (2011). "Don't let the Sport and Rec Officer get hold of it": Indigenous festivals, big aspirations and local knowledge. *Asia Pacific Journal of Arts and Cultural Management, 8*(1), 630–44.

Smajda, J., & Gerteis, J. (2012). Ethnic community and ethnic boundaries in a "sauce-scented neighborhood." *Sociological Forum, 27*(3), 617–40. http://dx.doi.org/10.1111/j.1573-7861.2012.01338.x

Small, K., Edwards, D., & Sheridan, L. (2005). A flexible framework for evaluating the socio-cultural impacts of a (small) festival. *International Journal of Event Management Research, 1*(1), 66–77.

Smallwood, C. (Ed.). (2010). *Librarians as Community Partners: An Outreach Handbook*. New York: American Library Association.

Smith, A. (2010). Leveraging benefits from major events: Maximising opportunities for peripheral urban areas. *Managing Leisure, 15*(3), 161–80. http://dx.doi.org/10.1080/13606710902752794

Smith, A. (2014). "Borrowing" public space to stage major events: The Greenwich Park controversy. *Urban Studies (Edinburgh, Scotland), 51*(2), 247–63. http://dx.doi.org/10.1177/0042098013489746

Smith, A.C., & Stewart, B. (2011). Organizational rituals: Features, function and mechanisms. *International Journal of Management Reviews, 13*(2), 113–33. http://dx.doi.org/10.1111/j.1468-2370.2010.00288.x

Smith, C.W. (2011). Staging auctions: Enabling exchange values to be contested and established. In B. Moeran & J.S. Pedersen (Eds.), *Negotiating Values in the Creative Industries: Fairs, Festivals and Competitive Events*, 94–118. New York: Cambridge University Press. http://dx.doi.org/10.1017/CBO9780511790393.005

Smith, J. (2014, 25 Aug.). Haitian, Brazilians celebrate together: Two cultures joined in Somerville festival with colorful art, music, food. *Boston Globe*, B2.

Smith, J.P. (1986). *Planning Community-Wide Special Events*. Fargo, ND: North Dakota State University Cooperative Extension Service.

Smith, J.W. (2012). Barriers and bridges to U.S. Forest Service – community relationships: Results from two pilot tests of a rapid social capital assessment protocol. *Forests, 3*(4), 1157–79. http://dx.doi.org/10.3390/f3041157

Smith, K. (2013). Portuguese in Toronto: Interplay between present, past, and future. *InterDISCIPLINARY Journal of Portuguese Diaspora Studies, 2*(1), 79–91.

Smith, L.S. (2001). Health of America's newcomers. *Journal of Community Health Nursing, 18*(1), 53–68. http://dx.doi.org/10.1207/S15327655JCHN1801_05

Smith, S., Bellaby, P., & Lindsay, S.L. (2010). Social inclusion at different scales in the urban environment: Locating the community to empower. *Urban Studies, 47*(7), 1439–57. http://dx.doi.org/10.1177/0042098009353618

Smith-Shank, D.L. (2002). Community celebrations as ritual signifiers. *Visual Arts Research, 28*(2), 57–63.

Snow, D.A., & Owens, P.B. (2013). *Crowds (Gatherings) and Collective Behavior (Action)*. Oxford: Blackwell. http://dx.doi.org/10.1002/9780470674871.wbespm462

Soja, E.W. (2010). *Seeking Social Justice*. Minneapolis, MN: University of Minnesota Press.

Solís, J., Fernández, J.S., & Alcalá, L. (2013). Mexican immigrant children and youth's contributions to a community centro. *Sociological Studies of Children and Youth, 16*, 177–200. http://dx.doi.org/10.1108/S1537-4661(2013)0000016012

Song, L.K. (2015). Race, transformative planning, and the just city. *Planning Theory, 14*(2), 152–73.

Sonn, C.C., & Quayle, A.F. (2013). Developing praxis: Mobilising critical race theory in community cultural development. *Journal of Community & Applied Social Psychology, 23*(5), 435–48. http://dx.doi.org/10.1002/casp.2145

Soomaroo, L., & Murray, V. (2012). Disasters at mass gatherings: Lessons from history. *PLoS Currents, 4*.

Speiser, V.M. (2014). Working in a troubled land: Using the applied arts towards conflict transformation and community health in Israel. *Journal of Applied Arts & Health, 4*(3), 325–34. http://dx.doi.org/10.1386/jaah.4.3.325_1

Spracklen, K., Richter, A., & Spracklen, B. (2013). The eventization of leisure and the strange death of alternative Leeds. *City: Analysis of Urban Trends, Culture, Theory, Policy Action, 17*(2), 164–78.

Stadler, R., Fullagar, S., & Reid, S. (2014). The professionalization of festival organizations: A relational approach to knowledge management. *Event Management, 18*(1), 39–52. http://dx.doi.org/10.3727/152599514X13883555341841

Stadler, R., Reid, S., & Fullagar, S. (2013). An ethnographic exploration of knowledge practices within Queensland Music Festival. *International Journal of Event and Festival Management, 4*(2), 90–106. http://dx.doi.org/10.1108/17582951311325872

Stanger-Ross, J. (2006). An inviting parish: Community without locality in postwar Italian Toronto. *Canadian Historical Review, 87*(3), 381–407.

Stanger-Ross, J. (2010). *Staying Italian: Urban Change and Ethnic Life in Postwar Toronto and Philadelphia*. Chicago: University of Chicago Press.

Stavrides, S. (2013). Contested urban rhythms: From the industrial city to the post-industrial urban archipelago. *Sociological Review, 61*(1), 34–50. http://dx.doi.org/10.1111/1467-954X.12052

Steele, J.S. (2013). Geopathology on May Day: Expressions of culture on Hawai'i's elementary school stages. *Equity & Excellence in Education, 46*(2), 169–83. http://dx.doi.org/10.1080/10665684.2013.779163

Steffensmeier, T. (2010). Building a public square: An analysis of community narratives. *Community Development, 41*(2), 255–68. http://dx.doi.org/10.1080/15575330903477317

Stern, M.J., & Seifert, S.C. (2010). Cultural clusters: The implications of cultural assets agglomeration for neighborhood revitalization. *Journal of Planning Education and Research, 29*(3), 262–79. http://dx.doi.org/10.1177/0739456X09358555

Stevens, Q., & Shin, H. (2012). Urban festivals and local social space. *Planning Practice and Research, 29*(1), 1–20.

Stockinger, P. (Ed.). (2013). *Digital Audiovisual Archives*. New York: Wiley. http://dx.doi.org/10.1002/9781118561980

Stodolska, M., Shinew, K.J., Acevedo, J.C. & Izenstark, D. (2011). Perceptions of urban parks as havens and contested terrains by Mexican-Americans in Chicago neighborhoods. *Leisure Sciences: An Interdisciplinary Journal, 33*(2), 103–26.

Stoecker, R., Tryon, E.A., & Hilgendorf, A. (Eds.). (2009). *The Unheard Voices: Community Organizations and Service Learning*. Philadelphia, PA: Temple University Press.

Stohl, C. (2014). Crowds, clouds, and community. *Journal of Communication, 64*(1), 1–19. http://dx.doi.org/10.1111/jcom.12075

Stone, C., & Millan, A. (2011). Empowering communities in the big society: Voluntarism and event management issues at the Cheetham Hill Cross-Cultural Festival. *International Journal of Management Cases, 13*(3), 242–50. http://dx.doi.org/10.5848/APBJ.2011.00059

Stover, J.A. (2013). Framing social movements through documentary films. *Contexts, 12*(4), 56–8. http://dx.doi.org/10.1177/1536504213511218

Strunk, C. (2014). "We are always thinking of our community": Bolivian hometown associations, networks of reciprocity, and indigeneity in Washington, DC. *Journal of Ethnic and Migration Studies, 40*(11), 1697–715.

Sullivan, G. B. (2014). Collective pride, happiness, and celebratory emotions: Aggregative, network, and cultural models. In C. von Scheve & M. Salmela (Eds.), *Collective Emotions*, 266–80. Oxford: Oxford University Press.

Sumner, J., Mair, H., & Nelson, E. (2010). Putting the culture back into agriculture: Civic engagement, community and the celebration of local food. *International Journal of Agricultural Sustainability, 8*(1), 54–61. http://dx.doi.org/10.3763/ijas.2009.0454

Sussman, N., & Barnes, T. (2014, 3 Mar.). At carnival, where challenging normal is the norm. *New York Times*, A11.

Swain, J., French, S., Barnes, C., & Thomas, C. (Eds.). (2013). *Disabling Barriers – Enabling Environments*. Thousand Oaks, CA: Sage.

Swan, P., & Atkinson, S. (2012). Managing evaluation: A community arts organisation's perspective. *Arts & Health: An International Journal for Research, Policy and Practice, 4*(3), 217–29. http://dx.doi.org/10.1080/17533015.2012.665372

Takaki, R. (2012). *Strangers from a Different Shore: A History of Asian Americans*. Updated and rev. Boston, MA: Little, Brown and Company.

Taks, M., Green, B.C., Chalip, L., Kesenne, S., & Martyn, S. (2013). Visitor composition and event-related spending. *International Journal of Culture, Tourism and Hospitality Research, 7*(2), 132–47.

Tallon, A. (2014). Festivals, carnivals and urban regeneration. *Journal of Urban Regeneration and Renewal, 7*(4), 301–6.

Tan, N.X., Tan, G.X., Yang, L.-G., Yang, B., Powers, K.A., Emch, M.E., & Tucker, J.D. (2014). Temporal trends in syphilis and gonorrhea incidences in Guangdong Province, China. *Journal of Infectious Diseases*, *209*(3), 426–30. http://dx.doi.org/10.1093/infdis/jit496

Teixeira, C. (2006). Residential segregation and ethnic economies in a multicultural city: The little Portugal of Toronto. In D.H. Kaplan & W. Li (Eds.), *Landscapes of the Ethnic Economy*, 49–65. New York: Rowman & Littlefield

Teixeira, C. (2007). Toronto's Little Portugal: A neighbourhood in transition. *Centre for Urban and Community Studies Research Bulletin*, *35*, 1–8.

Teixeira, C., Lo, L., & Truelove, M. (2007). Immigrant entrepreneurship, institutional discrimination, and implications for public policy: a case study in Toronto. *Environment and Planning. C, Government & Policy*, *25*(2), 176–93. http://dx.doi.org/10.1068/c18r

Tennant, M. (2013). Fun and fundraising: The selling of charity in New Zealand's past. *Social History*, *38*(1), 46–65. http://dx.doi.org/10.1080/03071022.2013.755390

Terry, N., Macy, A., & Owens, J.K. (2009). Bikers, aliens, and movie stars: Comparing the economic impact of special events. *Journal of Business & Economics Research*, *7*(11), 73–80.

Theodossopoulos, D. (2013). Emberá indigenous tourism and the trap of authenticity: Beyond inauthenticity and invention. *Anthropological Quarterly*, *86*(2), 397–425. http://dx.doi.org/10.1353/anq.2013.0023

Thibaud, J.-P. (2011). The three dynamics of urban ambiances. In B. LaBelle & C. Martinho (Eds.), *Sites of sounds: Of architecture and the ear*, vol. 2, 43–53. Berlin: Errant Bodies Press.

Thomas, T., & Kim, Y.H. (2011). A study of attendees' motivations: Oxford Film Festival. URJHS: Undergraduate Research Journal for the Human Sciences, *10*. http://www.kon.org/urc/urc_research_journal10.html

Thompson, D. (2011). Art fairs: The market as medium. In B. Moeran & J.S. Pedersen (Eds.), *Negotiating Values in the Creative Industries: Fairs, Festivals and Competitive Events*, 59–72. New York: Cambridge University Press. http://dx.doi.org/10.1017/CBO9780511790393.003

Throsby, D. (1999). Cultural capital. *Journal of Cultural Economics*, *23*(1/2), 3–12. http://dx.doi.org/10.1023/A:1007543313370

Tisdall, E.K.M. (2013). The transformation of participation? Exploring the potential of "Transformative Participation" for theory and practice around children and young people's participation. *Global Studies of Childhood*, *3*(2), 183–93. http://dx.doi.org/10.2304/gsch.2013.3.2.183

Tkaczynski, A. (2013a). Flower power? Activity preferences of residents and tourists to an Australian flower festival. *Tourism Analysis, 18*(5), 607–13. http://dx.doi.org/10.3727/108354213X13782245307993

Tkaczynski, A. (2013b). Festival performance (FESTPERF) revisited: Service quality and special events. In J. Chen (Ed.), *Advances in Hospital and Leisure*, 227–35. Bingley, UK: Emerald Group. http://dx.doi.org/10.1108/S1745-3542(2013)0000009015

Tkaczynski, A. (2013c). A stakeholder approach to attendee segmentation: A case study of an Australian Christian music festival. *Event Management, 17*(3), 283–98. http://dx.doi.org/10.3727/152599513X13708863377999

Tkaczynski, A., & Rundle-Thiele, S.R. (2011). Event segmentation: A review and research agenda. *Tourism Management, 32*(2), 426–34. http://dx.doi.org/10.1016/j.tourman.2010.03.010

Tkaczynski, A., & Rundle-Thiele, S.R. (2013). Understanding what really motivates attendance: A music festival segmentation study. *Journal of Travel & Tourism Marketing, 30*(6), 610–23. http://dx.doi.org/10.1080/10548408.2013.810998

Tovar, F.J., Arnal, M., Castro, C.D., Lahera-Sánchez, A., & Revilla, J.C. (2011). A tale of two cities: Working class identity, industrial relations and community in declining textile and shoe industries in Spain. *International Journal of Heritage Studies, 17*(4), 331–43. http://dx.doi.org/10.1080/13527258.2011.577966

Traverso-Yepez, M., Maddalena, V., Bavington, W. & Donovan, C. (2012, May). Community capacity building for health: A critical look at the practical implications of the approach. *Sage Open.*

Trinidad, A.M.O. (2012). Critical indigenous pedagogy of place: A framework to indigenize a youth food justice movement. *Journal of Indigenous Social Development, 1*(1), 1–17.

Tsai, J.-H., Yeh, S.-S., & Huan, T.-C. (2011), Creating loyalty by involvement among festival goers. In Joseph S. Chen (Ed.), *Advances in Hospitality and Leisure*, 173–91. Bingley, UK: Emerald Group.

Tucker-Raymond, E., Rosario-Ramos, E.M., & Rosario, M.L. (2011). Cultural persistence, political resistance, and hope in the community and school-based art of a Puerto Rican diaspora neighborhood. *Equity & Excellence in Education, 44*(2), 270–86. http://dx.doi.org/10.1080/10665684.2011.563678

Tull, J. (2012). Gathering festival statistics: The theoretical platforms and their relevance to building a global rubric. *Journal of Eastern Caribbean Studies, 37*(3/4), 40–70.

Tumwesigye, N.M., Atuyambe, L., Wanyenze, R.K., Kibira, S.P., Li, Q., Wabwire-Mangen, F., & Wagner, G. (2012). Alcohol consumption and risky sexual behaviour in the fishing communities: Evidence from two fish landing sites on Lake Victoria in Uganda. *BMC Public Health, 12*(1), 1069. http://dx.doi.org/10.1186/1471-2458-12-1069

Ulldemolins, J.R. (2014). Culture and authenticity in urban regeneration processes: Place branding in central Barcelona. *Urban Studies (Edinburgh, Scotland), 51*(14), 3026–45.

United Nations. (2013). *The Current Urbanization Prospects: The 2011 Revision.* New York: Department of Economic and Social Affairs, United Nations. http://www.un.org/en/development/desa/population/publications/pdf/urbanization/WUP2011_Report.pdf

U.S. Bureau of the Census. (2012). *The Asian population: 2010.* 2010 Census Briefs. Washington, D.C.: U.S. Department of Commerce.

Vaccari, A., Martino, M., Rojas, F., & Ratti, C. (2010). *Pulse of the City: Visualizing Urban Dynamics of Special Events.* Cambridge, MA: MIT Press.

de Valck, M., & Soeteman, M. (2010). "And the winner is …": What happens behind the scenes of film festival competitions. *International Journal of Cultural Studies, 13*(3), 290–307. http://dx.doi.org/10.1177/1367877909359735

Van Aalst, I., & van Melik, R. (2012). City festivals and urban development: Does place matter? *European Urban and Regional Studies, 19*(2), 195–206. http://dx.doi.org/10.1177/0969776411428746

van Campen, J.S., van Diessen, E., Otte, W.M., Joels, M., Jansen, F.E., & Braun, K.P.J. (2014). Does Saint Nicholas provoke seizures? Hints from Google trends. *Epilepsy & Behavior, 32*(2), 132–4. http://dx.doi.org/10.1016/j.yebeh.2014.01.019

van Heerden, E. (2011). The social and spatial construction of two South African arts festivals as liminal events. *South African Theatre Journal, 25*(1), 54–71. http://dx.doi.org/10.1080/10137548.2011.619723

Van Winkle, C.M., & Woosnam, K.M. (2014). Sense of community and perceptions of festival social impacts. *International Journal of Event and Festival Management, 5*(1), 22–38. http://dx.doi.org/10.1108/IJEFM-01-2013-0002

Van Winkle, C.M., Woosnam, K.M., & Mohammed, A.M. (2013). Sense of community and festival attendance. *Event Management, 17*(2), 155–63. http://dx.doi.org/10.3727/152599513X13668224082468

Vanderwaeren, E. (2014). Integrating by means of art? Expressions of cultural hybridisations in the city of Antwerp. *Identities: Global Studies in Culture and Power, 21*(1), 60–74. http://dx.doi.org/10.1080/1070289X.2013.846858

Vargas-Hernandez, J.G. (2012). Sustainable cultural and heritage tourism in regional development of Southern Jalisco. *World Journal of Entrepreneurship, Management and Sustainable Development, 8*(2–3), 146–61.

Veronis, L. (2006). The Canadian Hispanic Day Parade, or how Latin American immigrants practice (sub) urban citizenship in Toronto. *Environment & Planning, 38*(9), 1653–71. http://dx.doi.org/10.1068/a37413

Veronis, L. (2007). Strategic spatial essentialism: Latin Americans' real and imagined geographies of belonging in Toronto. *Social & Cultural Geography, 8*(3), 455–73. http://dx.doi.org/10.1080/14649360701488997

Veronis, L. (2010). Immigrant participation in the transnational era: Latin Americans' experiences with collective organising in Toronto. *Journal of International Migration and Integration/Revue de l'integration et de la migration internationale, 11*(2), 173–92.

Verschelden, G., Eeghem, E.V., Steel, R., de Visscher, S., & Dekeyrel, C. (2012). Positioning community art practices in urban cracks. *International Journal of Lifelong Education, 31*(3), 277–91. http://dx.doi.org/10.1080/02601370.2012.683607

Vieira, R.S., & Antunes, P. (2014). Using photo-surveys to inform participatory urban planning processes: Lessons from practice. *Land Use Policy, 38*(4), 497–508. http://dx.doi.org/10.1016/j.landusepol.2013.12.012

Villa, R.H. (2000). *Space and Place in Urban Chicano Literature and Culture*. Austin, TX: University of Texas Press.

Volker, S., & Kistemann, T. (2013). "I'm always entirely happy when I'm here!" Urban blue enhancing human health and well-being in Cologne and Dusseldorf, Germany. *Social Science & Medicine, 91*(2), 141–52. http://dx.doi.org/10.1016/j.socscimed.2013.04.016

Vukov, N. (2012). Cities, memorial sites, memory: The case of Plovdiv. *Our Europe: Ethnography-Ethnology-Anthropology of Culture, 2*, 129–44.

Waitt, G. (2008). Urban festivals: Geographies of hype, helplessness and hope. *Geography Compass, 2*(2), 513–37. http://dx.doi.org/10.1111/j.1749-8198.2007.00089.x

Waitt, G., & Stapel, C. (2013). "Fornicating on floats"? The cultural politics of the Sydney Mardi Gras Parade beyond the metropolis. *Leisure Studies, 30*(20), 197–216.

Wakefield, M.A., Loken, B., & Hornik, R.C. (2010). Use of mass media campaigns to change health behaviour. *Lancet, 376*(9748), 1261–71. http://dx.doi.org/10.1016/S0140-6736(10)60809-4

Wakimoto, D.K., Bruce, C., & Partridge, H. (2013). Archivist as activist: Lessons from three queer community archives in California. *Archival Science, 13*(4), 293–316. http://dx.doi.org/10.1007/s10502-013-9201-1

Waldinger, R. (2013). Engaging from abroad: The sociology of emigrant politics. *Migration Studies, 2*(3), 319–39.

Wardrop, K.M., & Robertson, M. (2004). Edinburgh's Winter Festival. In I. Yeoman, M. Robertson, J. Ali-Knight, S. Drummond, & U. McMahan-Beattie (Eds.), *Festival and Events Management: An International Arts and Culture Perspective*, 346–57. Oxford: Butterworth-Heinemann. http://dx.doi.org/10.1016/B978-0-7506-5872-0.50027-2

Walker, D., & Polepeddi, P. (2013). Becoming a multicultural services library: A guided journey to serving diverse populations. In C. Smallwood & K. Becnel (Eds.), *Library Services for Multicultural Patrons: Strategies to Encourage Library Use*, 3–12. Lanham, MD: Rowman & Littlefield.

Wan, Y.K.P., & Chan, S.H.J. (2013). Factors that affect the level of tourists' satisfaction and loyalty towards food festivals: A case study of Macau. *International Journal of Tourism Research*, 15(3), 226–40. http://dx.doi.org/10.1002/jtr.1863

Wang, C.C., Yi, W.K., Tao, Z.W., & Carovano, K. (1998). Photovoice as a participatory health promotion strategy. *Health Promotion International*, 13(1), 75–86. http://dx.doi.org/10.1093/heapro/13.1.75

Wang, J., Miller, N.A., Hufstader, M.A., & Bian, Y. (2007). The health status of Asian Americans and Pacific Islanders and their access to health services. *Social Work in Public Health*, 23(1), 15–43. http://dx.doi.org/10.1300/J523v23n01_02

Wang, T. (2012). Ritual meaning and recognition. In D. Solomon, R. Fan, & P.-C. Lu (Eds.), *Ritual and the Moral Life*, 89–104. New York: Springer.

Wardle, C., & West, E. (2004). The press as agents of nationalism in the Queen's Golden Jubilee: How British newspapers celebrated a media event. *European Journal of Communication*, 19(2), 195–214. http://dx.doi.org/10.1177/0267323104042910

Wasik, T. (2012). "Lower Manhattan medical centers." In CUNY's Professor Margaret Chin Macauley's Honors College Seminar Two, "Four Diverse Communities." http://macaulay.cuny.edu/eportfolios/mchin2012/579-2

Waters, R.D., Tindall, N.T., & Morton, T.S. (2010). Media catching and the journalist – public relations practitioner relationship: How social media are changing the practice of media relations. *Journal of Public Relations Research*, 22(3), 241–64. http://dx.doi.org/10.1080/10627261003799202

Waterton, E., & Smith, L. (2010). The recognition and misrecognition of community heritage. *International Journal of Heritage Studies*, 16(1–2), 4–15. http://dx.doi.org/10.1080/13527250903441671

Watson, R., & Yip, P. (2011). How many were there when it mattered? Estimating the sizes of crowds. *Significance*, 8(1), 104–7.

Weil, M., Gamble, D.N., & Ohmer, M.L. (2013). Evolution, models, and the changing context of community practice. In M. Weil, M. Reisch, &

M.L. Ohmer (Eds.), *The Handbook of Community Practice*, 167–93. Los Angeles, CA: Sage.

Weller, S.A. (2013). Consuming the city: Public fashion festivals and the participatory economies of urban spaces in Melbourne, Australia. *Urban Studies*, 50(14), 2853–68. http://dx.doi.org/10.1177/0042098013482500

Welty Peachey, J., Borland, J., Lobpries, J., & Cohen, A. (2014). Managing impact: Leveraging sacred spaces and community celebration to maximize social capital at a sport-for-development event. *Sport Management Review*, 18(1), 86–98.

Wen, B., Yang, Z., & Huang, Q. (2013, June). Research on volunteer service system construction of city major festival activity: A case of the Asia-Europe Exposition of Urumqi. In X.M. Huang (Ed.), *Proceedings of the 2013 the International Conference on Education Technology and Information System (ICETIS 2013)*, 156–60. Amsterdam: Atlantis Press.

West, G.R. (2012). *Creating Community: Finding Meaning in the Place We Live.* Bloomington, IN: Canadian Institute of Cultural Affairs.

Wexler, L. (2014). Looking across three generations of Alaska Natives to explore how culture fosters indigenous resilience. *Transcultural Psychiatry*, 51(1), 73–92. http://dx.doi.org/10.1177/1363461513497417

Whalen, P. (2010). "The return of Crazy Mother": The cultural politics of carnival in 1930s Dijon. *Social Identities: Journal for the Study of Race, Nature and Culture*, 16(4), 471–96.

Wherry, F.F. (2011). *The Philadelphia Barrio: The Arts, Branding, and Neighborhood Transformation*. Chicago: University of Chicago Press.

Whitford, M., & Ruhanen, L. (2013). Indigenous festivals and community development: A sociocultural analysis of an Australian indigenous festival. *Event Management*, 17(1), 49–61.

White, J. (2012). The immigration experience: Losses and gains for immigrant and refugee women. In L. Williams, R. Roberts, & A. McIntosh (Eds.), *Radical Human Ecology: Intercultural and Indigenous Approaches*, 291–312. Surrey: Ashgate.

Whitesel, J., & Shuman, A. (2013). Normalizing desire: Stigma and the carnivalesque in gay bigmen's cultural practices. *Men and Masculinities*, 16(4), 478–496. http://dx.doi.org/10.1177/1097184X13502668

Wilks, L. (2011). Social capital in the music festival experience. In S. Page & J. Connell (Eds.), *The Routledge Handbook of Events*, 260–72. London: Routledge.

Williams, M., & Bowdin, G.A.J. (2010). Festival evaluation: An exploration of seven UK arts festivals. *Managing Leisure*, 12(2–3), 187–203.

Williams, R. (2012). *We Been Here, We Live Here, We Love Here: Black Lesbians' Performance of Presence in Chicago's Southside*. Evanston, IL: Northwestern University.

Williams, V., & Culp, G.L. (2011). *Storage*. London: University of the Arts.

Wilson, E. (2013). *Foodborne Illnesses and Seasonality Related to Mobile Food Sources and Festivals in Georgia*. Walden University Dissertation.

Wilson-Forsberg, S. (2014). "We don't integrate; we adapt:" Latin American immigrants interpret their Canadian employment experiences in Southwestern Ontario. *Journal of International Migration and Integration, 16*(3), 469–89.

Wintemute, G.J., Claire, B., McHenry, V., & Wright, M.A. (2011). Stray bullet shootings in the United States. *Journal of the American Medical Association, 306*(5), 491–2. http://dx.doi.org/10.1001/jama.2011.1066

Winter, R., Fraser, S., Booker, N., & Treloar, C. (2013). Authenticity and diversity: Enhancing Australian hepatitis C prevention messages. *Contemporary Drug Problems, 40*(4), 505–29. http://dx.doi.org/10.1177/009145091304000404

Wirz, M., Franke, T., Roggen, D., Mitleton-Kelly, E., Lukowicz, P., & Troster, G. (2012). Inferring crowd conditions from pedestrians' location traces for real-time crowd monitoring during city-scale mass gatherings. In S. Reddy & K. Drira (Eds.), *Proceedings 21st IEEE WETICE Conference (WETICE 2012), 25–27 June, Toulouse, France*, 367–372. New York: IEEE.

Wirz, M., Mitleton-Kelly, E., Franke, T., Camilleri, V., Montebello, M., Roggen, D., Lukowicz, P., & Troster, G. (2013). Using mobile technology and a participatory sensing approach for crowd monitoring and management during large-scale mass gatherings. In E. Mitleton-Kelley (Ed.), *Co-evolution of Intelligent Socio-technical Systems*, 61–77. Berlin: Springer. http://dx.doi.org/10.1007/978-3-642-36614-7_4

Wolch, J., Newell, J., Seymour, M., Huang, H.B., Reynolds, K., & Mapes, J. (2010). The forgotten and the future: Reclaiming back alleys for a sustainable city. *Environment & Planning A, 42*(12), 2874–96. http://dx.doi.org/10.1068/a42259

Wong, B.P. (1998). *Ethnicity and Entrepreneurship: The New Chinese Immigrants in the San Francisco Bay Area*. Boston: Allyn and Bacon.

Wong, J., Wu, H.C., & Cheng, C.C. (2014). An empirical analysis of synthesizing the effects of festival quality, emotion, festival image and festival satisfaction on festival loyalty: A case study of Macau Food Festival. *International Journal of Tourism Research*. http://dx.doi.org/10.1002/jtr.2011

Wood, E.H. (2009). Evaluating event marketing: Experience or outcome? *Journal of Promotion Management, 15*(1–2), 247–68. http://dx.doi.org/10.1080/10496490902892580

Ward, L. (2008). *Fair, Festival, & Event Promotion Handbook*. http://festivalsandevents.com/festival-planning/

Wronska-Friend, M. (2012). "Why haven't we been taught all that at school?" Crosscultural community projects in North Queensland, Australia. *Curator: The Museum Journal, 55*(1), 3–19. http://dx.doi.org/10.1111/j.2151-6952.2011.00117.x

Wu, M.-Y. (2014). Approaching tourism: Perspectives from the young hosts in a rural heritage community in Tibet. *Current Issues in Tourism.* http://dx.doi.org/10.1080/13683500.2014.889091

Wu, M.-Y., & Pearce, P.L. (2013a). Tourists to Lhasa, Tibet: How local youth classify, understand and respond to different types of travelers. *Asia Pacific Journal of Tourism Research, 18*(6), 549–72. http://dx.doi.org/10.1080/10941665.2012.680975

Wu, M.-Y., & Pearce, P.L. (2013b). Asset-based community development as applied to tourism in Tibet. *Tourism Geographies: An International Journal of Tourism Space, Place and Environment, 16*(3), 438–456.

Wu, W.C., Chen, W.X., & Huan, T.C. (2010). Evaluating performance factors of art festivals: A case study in Taiwan. *Advances in Hospitality and Leisure, 6,* 99–115. http://dx.doi.org/10.1108/S1745-3542(2010)0000006010

Wu, Y. (2014). *Principles of Cultural Competency and the Implications for Western Evaluators Using the Program Evaluation Standards in Chinese Cultural Context.* Queen's University, Rosa Bruno-Jofré Symposium in Education.

Wyatt, K. (2014, 2 Aug.). Denver county fair adds contests for marijuana entries. *Boston Globe,* A6.

Xu, N. (2013). *Why Chinatown Has Gentrifed Later than Other Comunities in Downtown Manhattan: A Planning History.* Doctoral dissertation, Columbia University.

Yang, J., Gu, Y., & Cen, J. (2011). Festival tourists' emotion, perceived value, and behavioral intentions: A test of the moderating effect of festivalscape. *Journal of Convention & Event Tourism, 12*(1), 25–44. http://dx.doi.org/10.1080/15470148.2010.551292

Yang, L., Gao, X., Wang, X., Nie, W., Wang, J., Gao, R., ..., & Wang, W. (2014). Impacts of firecracker burning on aerosol chemical characteristics and human health risk levels during the Chinese New Year celebration in Jinan, China. *Science of the Total Environment, 476*(1), 57–64. http://dx.doi.org/10.1016/j.scitotenv.2013.12.110

Yang, L., & Wall, G. (2009). Authenticity in ethnic tourism: Domestic tourists' perspectives. *Current Issues in Tourism, 12*(3), 235–54. http://dx.doi.org/10.1080/13683500802406880

Yang, Z.Z., & Hu, H.X. (2011). Criticism about tourism authenticity. *Tourism Tribune.* http://en.cnki.com.cn/Article_en/CJFDTOTAL-LYXK201112020.htm

Yao, S.-J., & Heng, C.K. (2013). An (extra)ordinary night out: Urban informality, social sustainability and the night-time economy. *Urban Studies*, *52*(1), 606–16.

Yee, V. (2014, 30 June). With rainbow neckerchiefs, celebrating pride and progress at parade. *New York Times*, A17.

Yee, V., & Turkewitz, J. (2014, 9 June). After scandal, the Puerto Rican Day Parade goes "back to its roots." *New York Times*, A12.

Yeh, C. (2002). Contesting identities: Youth rebellion in San Francisco's Chinese New Year festivals, 1953–1969. In S.L. Cassel (Ed.), *The Chinese In America: A History from Gold Mountain to the New Millennium*, 329–50. Walnut Creek, CA: AltaMira Press.

Yeh, C. (2004). "In the traditions of China and in the freedom of America": The making of San Francisco's Chinese New Year festivals. *American Quarterly*, *56*(2), 395–420. http://dx.doi.org/10.1353/aq.2004.0029

Yeh, C. (2008). *Making an American Festival: Chinese New Year in San Francisco's Chinatown*. Berkeley, CA: University of California Press.

Yeh, C. (2009). Politicizing Chinese New Year festivals: Cold war politics, transnational conflicts and Chinese America. In M.H. Ross (Ed.), *Culture and Belonging in Divided Societies: Contestation and Symbolic Landscapes*, 238–58. Philadelphia: University of Pennsylvania Press.

Yeh, S.-L. (2013). Pig sacrifices, mobility and the ritual recreation of community among the Amis of Taiwan. *Asia Pacific Journal of Anthropology*, *14*(1), 41–56. http://dx.doi.org/10.1080/14442213.2012.747557

Yin, R.K. (2014). *Case Study Research: Design and Methods*. Thousand Oaks, CA: Sage.

Yogev, T., & Grund, T. (2012). Network dynamics and market structure: The case of art fairs. *Sociological Focus*, *45*(1), 23–40. http://dx.doi.org/10.1080/00380237.2012.630846

Yolal, M., Cetinel, P., & Uysal, M. (2009). An examination of festival motivation and perceived benefits relationship: Eskisehir International Festival, *Journal of Convention & Event Tourism 10*(4), 276–91.

Yoo, I.Y., Lee, T.J., & Lee, C.K. (2015). Effect of health and wellness values on festival visit motivation. *Asia Pacific Journal of Tourism Research*, *20*(2), 152–70.

Yoon, Y.-S., Lee, J.-S., & Lee, C.-K. (2010). Measuring festival quality and value affecting visitors' satisfaction and loyalty using a structural approach. *International Journal of Hospitality Management*, *29*(2), 335–42. http://dx.doi.org/10.1016/j.ijhm.2009.10.002

Young, J.R. (2012). *It Takes a Community: Civic Life and Community Involvement among Coös County Youth.* http://scholars.unh.edu/carsey/178

Yung, J. (2006). *San Francisco's Chinatown*. San Francisco: Arcadia.

Yuval-Davis, N. (2006). Belonging and the politics of belonging. *Patterns of Prejudice*, *40*(3), 197–214. http://dx.doi.org/10.1080/00313220600769331

Zahra, A., & McGehee, N.G. (2013). Volunteer tourism: A host community capital perspective. *Annals of Tourism Research*, *42*(1), 22–45. http://dx.doi.org/10.1016/j.annals.2013.01.008

Zapata-Barrero, R. (2014). The limits of shaping diversity as public culture: Permanent festivities in Barcelona. *Cities*, *37*(1), 66–72. http://dx.doi.org/10.1016/j.cities.2013.11.007

Zerubavel, E. (2012). *Time Maps: Collective Memory and the Social Shape of the Past*. Chicago: University of Chicago Press.

Zhang, X.L., Weng, W.G., Yuan, H.Y., & Chen, J.G. (2013). Empirical study of a unidirectional dense crowd during a real mass event. *Physica A: Statistical Mechanics and its Applications*, *392*(12), 2781–91.

Zhou, M. (2010). *Chinatown: The Socioeconomic Potential of an Urban Enclave*. Philadelphia: Temple University Press.

Zhou, M., & Lee, R. (2013). Transnationalism and community building: Chinese immigrant organizations in the United States. *Annals of the Academy of Political and Social Science*, *647*(1), 22–49. http://dx.doi.org/10.1177/0002716212472456

Zhu, L. (2012). *National Holidays and Minority Festivals in Canadian Nation-Building*. Dissertation, University of Sheffield, UK.

Ziakas, V. (2010). Understanding an event portfolio: The uncovering of interrelationships, synergies, and leveraging opportunities. *Journal of Policy Research in Tourism, Leisure and Events*, *2*(2), 144–64. http://dx.doi.org/10.1080/19407963.2010.482274

Ziakas, V. (2013a). Fostering the social utility of events: An integrative framework for the strategic use of events in community development. *Current Issues in Tourism*. http://dx.doi.org/10.1080/13683500.2013.849664

Ziakas, V. (2013b). *Event Portfolio Planning and Management: A Holistic Approach*. New York: Routledge.

Ziakas, V., & Boukas, N. (2013). Extracting meanings of event tourist experiences: A phenomenological exploration of Limassol carnival. *Journal of Destination Marketing & Management*, *2*(2), 94–107. http://dx.doi.org/10.1016/j.jdmm.2013.02.002

Ziakas, V., & Costa, C.A. (2010a). "Between theatre and sport" in a rural event: Evolving unity and community development from the inside-out. *Journal of Sport & Tourism*, *15*(1), 7–26. http://dx.doi.org/10.1080/14775081003770892

Ziakas, V., & Costa, C.A. (2010b). Explicating inter-organizational linkages of a host community's events network. *International Journal of Event and Festival Management*, *1*(2), 132–47. http://dx.doi.org/10.1108/17852951011056919

Ziakas, V., & Costa, C.A. (2011). Event portfolio and multi-purpose development: Establishing the conceptual grounds. *Sport Management Review*, *14*(4), 409–23. http://dx.doi.org/10.1016/j.smr.2010.09.003

Ziakas, V., & Costa, C.A. (2012). "The show must go on": Event dramaturgy as consolidation of community. *Journal of Policy Research in Tourism, Leisure and Events*, *4*(1), 28–47. http://dx.doi.org/10.1080/19407963.2011.573392

Zigkolis, C., Papadopoulos, S., Filippou, G., Kompatsiaris, Y., & Vakali, A. (2014). Collaborative event annotation in tagged photo collections. *Multimedia Tools and Applications*, *70*(1), 89–118.

Zillinger, M. (2014). Media and the scaling of ritual spaces in Morocco. *Social Compass*, *61*(1), 39–47. http://dx.doi.org/10.1177/0037768613513941

Zinkhan, G.M., Fontenelle, S.D.M., & Balazs, A.L. (1999). The structure of Sao Paulo street markets: Evolving patterns of retail institutions. *Journal of Consumer Affairs*, *33*(1), 3–26. http://dx.doi.org/10.1111/j.1745-6606.1999.tb00758.x

Zuev, D., & Virchow, F. (2014). Performing national-identity: The many logics of producing national belonging in public rituals and events. *Nations and Nationalism*, *20*(2), 191–9. http://dx.doi.org/10.1111/nana.12063

Zukin, S. (2010). *The Naked City: The Death and Life of Authentic Urban Places*. New York: Oxford University Press.

Zukin, S. (2011). Reconstructing the authenticity of place. *Theory and Society*, *40*(2), 161–5. http://dx.doi.org/10.1007/s11186-010-9133-1

Index

academic institutions: assistance with celebrations, 7, 95–6; definitions of celebrations, 26–9, 47; service-learning projects, 106–7, 111, 121; terminology issues, 19, 26–9, 30, 40, 42–3, 47; training of practitioners, 186, 192–3

academic research: action research, 77–8; case studies, 79; conceptual frameworks, 17–18; critical epistemology, 191; definitions of celebrations, 26–9, 47; domains approach, 47; on economic capital, 32, 66, 74; ethnography, 10, 77–9, 190–1; on evaluation, 65, 68–9, 75; impact analysis, 65–6, 190; labeling and stereotypes, 93; and local knowledge, 95; multidisciplinary teams, 32–3, 40, 191; on national days, 164; need for more research, 13, 32–3; participatory methods, 91, 191; photography's use in, 77–8; ritual as framework, 40–2; terminology issues, 19, 26–9, 30, 40, 42–3, 47; unit of analysis, 191

African heritage, joint community events, 71–2

age groups: and acculturation, 181, 185; empowerment of, 68; at fairs, 103, 106–7, 111; history of the community, 181; as human capital, 22–3; intergenerational relationships, 23, 41–2, 87–8, 103, 107, 111, 188–9; motivation of attendees, 35, 36, 73–4; and rituals, 41–2; safety concerns, 55; segment exclusion, 76–7. *See also* children; older adults; youth and young adults

agricultural fairs, 103, 106

alcohol and illicit drugs: aggressive behaviors, 54–5, 122; marijuana fairs, 110; risky sexual behavior, 56; safety concerns, 55, 56; wine festivals, 117

ambiance: about, 37–8; evaluation of, 69, 77, 190; at fairs, 107; of successful events, 82

animals at events, 109

animation festivals, 117

anti-social behavior: aggressive behaviors, 54–5, 122; evaluation of, 72–3; at festivals, 58, 127; at parades, 59, 135–6; risky sexual behavior, 56. *See also* alcohol and illicit drugs; safety issues

archives. *See* records and archives
Arcodia, C.V., 30, 186
arts: about, 44–5; arts activism and social justice, 87, 92; capital/assets framework, 18–20, 105–6, 134; and community identity, 44; creative writing, 44–5; cultural competence and humility, 94–5; economic capital, 22, 86, 105–6, 134; evaluation using, 190; LGBTQ events, 173–4; murals and sculptures, 13, 106; parades, 86, 134, 173–4; performing arts, 22. *See also* cultural capital; music
art fairs and festivals: about, 105–6, 116–17; arts and crafts, 117; film festivals, 116, 117, 125–6; mental health arts festivals, 116; statistics on, 117; types of, 46, 103, 116–17
Asian communities: Chinatowns' histories, 144–6, 166; health fairs, 149. *See also entries beginning with* Chinese communities
assessment. *See* evaluation
assets, community. *See* community assets
assets/capital framework. *See* capital/assets framework
attendees: about, 33–6; attendee-to-attendee relationships, 52; diversity of, 35; documented status of, 32, 60, 74, 109, 151; evaluation of attendance, 67–8, 73–4, 125; at festivals, 22, 124; former residents, 34; gender, 35, 68, 74; gravitational force of events, 34–5, 81; loyalty of, 35, 52; motivation of, 33–6, 73–4, 121, 125; research needed on, 31, 33–4, 52; satisfaction of expectations of, 22, 73–4;

sexual identity, 35–6, 74; time of attendance, 126. *See also* crowds; motivation of attendees
auctions, 103
Australia: Chinese heritage, 22; community art, 80; Indigenous festivals, 117; LGBTQ events, 58
authenticity: about, 15, 42–4, 181; and cultural values, 15, 181; definition of, 42–3; and development goals, 15; and exploitation of cultural symbols, 15–16; literature on, 42–3

Bartleet, B.L., 29
beauty pageants, 159–62, 170
Belfast, parades, 35
Belgrade, Serbia, LGBTQ events, 129
Bendigo, Australia, Chinese heritage, 22
Berridge, G., 32
beverages. *See* food
bicycle protest parades, New York City, 131
birding festivals, 46
bisexual community. *See* LGBTQ communities
bling/dancehall funerals, Jamaica, 46
bonfires, 57
book fairs and festivals, 103, 117, 151–2
Boston, Massachusetts: Calling Music Festival, 108; Chinatown's history, 144–5; St Patrick's Day parades, 133, 136, 139–40
Boy Scouts and LGBTQ issues, 128
Bramadat, P.A., 166
branding of neighborhoods: about, 14–15; and authenticity, 43; and commercialism, 16; to counter

Index 279

racial tensions, 160; and cultural capital, 14–15, 22; and economic capital, 22; festivals, 116; and tourism, 14, 16
Brazil, fairs and carnivals, 108, 122
Brazilian communities, 71–2
Brown, S., 72, 96
Bud Billiken Day Parade, Chicago, 133
Bulgarian Society, Canada, 51
Burning Man Festival, 57
business and trade events, 38
businesses, small. See small businesses

California: Chinese communities, 144–5, 156–8; Rose Day Parade, 137. See also Chinese community, San Francisco
Calling Music Festival, Mass., 108
capacity enhancement: about, 10, 85–8, 96–8; community assets, 13–17, 85, 93–4; definition of, 85; excluded communities, 86; festivals, 120; frameworks for, 17–18, 84; literature on, 13, 16; media coverage and relations, 184–5; place and space, 86–8; practitioners' roles, 96–7; principles for, 96–8; research needed on, 52; rewards and benefits, 39–40; social change, 87; social inclusion, 17; values, 84–5, 88–9, 96–8. See also empowerment and ownership
capital/assets framework: about, 17–24, 84, 93–4, 96–8; case study on, 36; categorization of capital, 19; changes in, 18, 23; evaluation of, 68–9; and excluded communities, 71; frameworks, 17–18, 84; literature on, 18, 19; multiple forms of capital, 36; music and art festivals, 18–19; place and space, 19; principles and values, 96–8; resilience and culture, 18, 49–50; social anchors, 37; terminology, 19; types of capital, 19, 36; values and traditions, 84–5, 181–2; volunteers, 48. See also cultural capital; economic capital; human capital; intangible capital; physical capital; political capital; social capital
career and job fairs, 103–4, 108
carnivals, 115, 122. See also festivals
case studies: on evaluation, 79; fairs (Chinatown's Health Fair), 143–53, festivals (Chinese Lunar New Year Festival), 155–63; parades (Toronto events), 164–75
celebrations: about, 25–9, 188–9, 192–4; academic support for, 95–6; ambiance, 37–8, 69, 77, 82, 107; authenticity, 15–16, 42–4, 181; benefits, 39–40; boundaries in, 27–8; categorization of, 26, 29, 38–9, 47; collective memories, 6, 37, 49, 51; combined events, 26, 71–2, 87, 117, 119, 122, 158–9, 160, 164, 172; commercialization of, 15–16, 43–4, 193–4; definition used in this book, 28–9; definitions of, 26–9, 45, 47; disruption of everyday life, 6, 51; functions of, 6, 14; history of the community, 180–1; ideal prototype, 63; liminality of, 29–30; negative impacts of, 59–62; recent trends, 193–4; reports on, 82–3; as rituals, 40–2; sizes of, 28–9, 192; as social anchors, 37; sponsoring organizations, 193–4; stages in, 45, 97–8; terminology, 26–9; time span,

Index

25–6, 28, 32–3; types of, 38–9; and values, 84–5. *See also* fairs; festivals; goals, objectives, and purposes; parades

celebrations and practitioners. *See* practitioners' knowledge and skills; practitioners' roles

Charles B. Wang Community Health Center, New York City, 143–4, 146–53

Chicago, Illinois: Bud Billiken Day Parade, 133; Chinatown's history, 144–5; community gardens, 10; Latino parades, 132; Puerto Rican community, 10, 132; Turkish festivals, 128

The Child in the City (Ward), 17

children: about, 17, 90, 92; art festivals, 117; at fairs, 104, 106, 107–8; health and safety concerns, 55, 57, 58, 109; as human capital, 92; at LGBTQ events, 128–9; newcomers, 50, 51; noise as health issue, 57; participation of, 90, 92; and public spaces, 17, 118; venues for events, 118. *See also* schools

Chile, LGBTQ events, 129

China: emigration, 157–8; festivals, 57, 117, 155

Chinese communities: festivals, 155, 165; heritage assets, 22; historical background, 156–8, 160–1; organizations for newcomers, 50–1

Chinese community, New York City: case study (Chinatown's Health Fair, NYC), 143–53; Chinatown's history, 144–6; Chinese New Year celebrations, 57; festivals, 155; historical background, 156–8, 160; language barriers, 147, 151

Chinese community, San Francisco: case study (Lunar New Year Festival), 155–63; combined events, 158–62; Community Street Fair, 158–9, 162; diversity within, 161; historical background, 144–6, 155–8, 160; Lunar Day Parade, 158, 160–2; marketing, 161–2; Miss Chinatown beauty pageant, 159–62; New Year Flower Fair, 158, 162; New Year Run, 159; racial and political tensions, 159–60, 162–3; small businesses, 156; sociocultural background, 157, 159–60; sponsorships, 159–60; sustainability of festivals, 162; tourism, 157–63

Cinco de Mayo, 55–6, 115

cities. *See* communities

civic engagement. *See* volunteers and civic engagement

Cleveland, Ohio, community gardens, 10–11

cohesion of communities. *See* social capital

collaboration. *See* partnerships

collective memories, 6, 9, 21, 36, 37, 49, 51, 111. *See also* memory

Cologne, Germany, festival, 50–1

combined events, 26, 71–2, 87, 117, 119, 122, 158–60, 164, 172

commercialism: about, 55–6; and authenticity, 15–16, 43–4; and branding, 15–16; and cultural symbols, 15–16; and failed events, 60

communities: about, 9–12; assets, 13–17; authenticity, 42–4; community participation, 90–2; deficit paradigm, 7–9; definitions of, 9–10; diversity within, 35, 38, 43, 161; historical backgrounds, 80,

119, 180–1; insularity of, 147; social anchors in, 37; urban cracks, 37
community assets: about, 13–18, 93–4; categorization of, 18; literature on, 13, 16; resilience, 18, 49–50; synergistic relations, 31–2. *See also* capital/assets framework
community capacity enhancement. *See* capacity enhancement
community capital. *See* capital/assets framework
community gardens. *See* gardens, community
Community Social Work Practice in an Urban Setting (Delgado), 13
community social workers. *See* practitioners' knowledge and skills; practitioners' roles
community volunteers. *See* volunteers and civic engagement
contests and competitions, 45, 56, 101
corporate funding, 15, 55–6, 75–6, 189. *See also* funding and sponsorships
county and country fairs, 102–3, 104, 106, 110. *See also* fairs
craft fairs, 106
creative and competitive events, 26. *See also* celebrations
creative writing, 44–5
Creole Christmas, New Orleans, 127
Crompton, J.L., 35
crowds: about, 53–5; digital screens, 54; estimates of size, 54, 108; evaluation of, 75; at festivals, 154; and groups, 54; health system challenges, 57–8; literature on, 54, 57; management of, 54–5; at parades, 135; safety concerns, 54–8, 135, 154; social groups, 6, 20; stampedes, 57; time of attendance, 126; warmth and belonging, 20
cultural capital: about, 19–20, 31; and authenticity, 42–4; and branding of communities, 14–15; commercialization of, 15–16, 43–4, 193–4; community gardens, 87–8; and cultural artifacts, 22; cultural competence and humility, 94–5; development of, 21–2, 88; and economic capital, 19–20, 31, 158; evaluation of, 68–9; and festivals, 31, 119; literature on, 19, 51–2; as social anchor, 37; and transnationalism, 48–53. *See also* arts; authenticity; capital/assets framework

Dahl Handi, India, 47
dance festivals, 117
Day of the Dead, 46
deficit paradigm, 7–9
Delgado, M., 13, 96–7
Delphi technique, 81–2
democratic principles: about, 23, 193; and digital technologies, 77–8; and evaluation, 71; and participation, 71–2, 80; and political capital, 20; and segment exclusion, 76–7; and social media, 184–5; urban citizenship, 91–2; and urban cracks, 37
demographics: changes in, 52; and evaluation, 73–4; and gentrification, 52; history of the community, 180–1; and motivation of attendees, 35–6; newcomers, 52. *See also* age groups
Denmark, food festivals, 46
Denver, marijuana fair, 109

Detroit, Michigan, Halloween, 58
Devine, L., 69
Día de los Muertos, Latin America, 46
Dickson, G., 186
disabilities, people with: assets of, 59; inclusion in events, 59, 81; intersectionality with LGBTQ community, 173; labeling and stereotypes, 93; and photography, 81; and public spaces, 17; sales of art at fairs, 106; segment exclusion, 76–7
discriminatory practices. *See* excluded communities
documented status of participants, 32, 60, 74, 109, 151
documents. *See* records and archives
domains approach, 47
drinking and drug use. *See* alcohol and illicit drugs
Duisburg, Germany, parade, 135
Durkheim, Émile, 40, 125
Dyke March, 172

Eberle, S.G., 36
economic capital: about, 19–20, 30–2; and cultural capital, 19–20, 22, 31, 158; evaluation of, 31, 66, 68–9, 70, 72–4, 106, 123; fairs, 105–6; festivals, 19–20, 30–1, 119, 122, 123, 127, 159–60; and gravitational force of events, 34–5; informal economy, 31–2; literature on, 19, 30–2; negative impacts of, 70; and other forms of capital, 31; scholarship on, 48. *See also* capital/assets framework; partnerships; small businesses; tourism
Edinburgh Festival, 22

education and learning: about, 45; collective learning, 45; at fairs, 106, 111–12; as human capital, 22–3; intergenerational learning, 20, 22–3; partnerships, 111–12, 120; resilience and culture, 18; service-learning projects, 106–7, 111, 121. *See also* schools
Ehrenfeucht, R., 6, 27, 130
The Elementary Forms of the Religious Life (Durkheim), 125
employment fairs, 104
empowerment and ownership: about, 68, 92–3; and age, 68, 92; and authenticity, 42–3; community ownership, 30, 34, 85, 96–7, 182; decision making, 91; evaluation of, 68; and failed events, 60; health fairs, 146–7, 151; inclusion vs. participation, 91; local knowledge, 95; as organic process, 42; and privatization of public space, 14, 16, 60; research needed on, 34; and social justice, 92–3; sociodemographic factors, 35–6; and sponsorships, 75–6, 189; urban citizenship, 91–2; and values, 92–3. *See also* democratic principles; political tensions; volunteers and civic engagement
engagement, civic. *See* volunteers and civic engagement
environmental autobiographies, 9
environmental issues: about, 72; evaluation of, 72–3; at fairs, 107; at festivals, 126–7; political action in parades, 173; smoking, 57; water and sanitation concerns, 46, 47; wildlife festivals, 46
Epstein, W.M., 93

ethics and social issues. *See* social justice
ethnicity. *See* race and ethnicity
Ethnicity and Entrepreneurship (Wong), 156
ethnography, 77–9, 190–1
Eurochocolate Festival, Italy, 46
evaluation: about, 63–5, 72–3, 83, 189–91; and capacity enhancement, 65–6, 97–8; community-centered process, 64–73, 80; critical questions, 72–3, 191; and democratic participation, 71–2, 77–8; and excluded communities, 67, 71, 76–7, 82, 190–1; for funding and sponsors, 63, 67; literature on, 64–5, 69, 70; practitioners' roles, 65–6, 67, 75, 189–91; reports on, 82–3, 97; research needed on, 64–5; resources for, 70; rewards and benefits to community, 39–40, 66; as storytelling, 66, 69–70, 77–8; and values, 63, 65–6, 72
evaluation, areas: attendees, 34, 67–8; community characteristics, 19; economic impacts, 30–1, 64, 66, 68–9, 123, 137; empowerment and ownership, 68; fairs, 69, 110–11; festivals, 123–5; intangibles, 69, 123–4; management, 123; marketing, 67–8; motivation, 73–4; parades, 137; socio-cultural impacts, 64–6, 68–9, 72–4
evaluation, methods: about, 65–7, 69–70, 77, 190–1; ambiance approach, 69, 77, 190; case studies, 79; community-based research, 91; Delphi technique, 81–2;

ethnography, 77–9, 190–1; impact analysis, 65–6; interviews, 110–11; longitudinal studies, 72, 74, 75; mixed-method approaches, 70, 79; multidisciplinary teams, 191; participatory approaches, 80, 110–11, 137; photography, 77–8, 81, 137; qualitative/quantitative dimensions, 69–70, 80–2, 91, 110–11, 137, 190–1; team approach, 191; technologies, 80–1, 137; user-friendly methods, 75; visual approaches, 77–8
event medicine: challenges for, 57–8; at parades, 135; staff for health fairs, 152
event planners, professional, 60, 106–7, 179, 186
events and practitioners. *See* practitioners' knowledge and skills; practitioners' roles
excluded communities: about, 71–2, 76–7, 129, 190; ageism, 59; and evaluation, 67, 71, 76–7, 82, 190–1; at festivals, 120, 123, 129; inclusion strategies, 59, 91, 183; labeling and stereotypes, 93; multiple stigmas, 50; newcomers, 171; place and space, 86, 171; segregated communities, 7; and sustainability, 129; undocumented newcomers, 32, 60, 74, 109, 151. *See also* age groups; LGBTQ communities; race and ethnicity

failed events: about, 60–2; causes of, 60, 70, 122; evaluation of, 72–3; negative impacts of, 23, 59–61, 70; tips on how to fail, 61
fairs: about, 101–2, 112–13, 153; ambiance, 107; animals at, 109;

attendees, 105, 108; capital and assets, 105, 112–13; case studies on, 79; case study (NYC Chinatown's Health Fair), 143–53; challenges of, 108–10; classification of, 38–9, 47, 105; contests at, 101; county and country fairs, 102–3, 104, 110; environmental issues, 107; evaluation of, 69, 79, 110–11; food at, 101, 104, 106, 109; funding and sponsorships, 107; goals and objectives, 101–2, 112–13; green spaces, 88; human services, 101, 104, 105, 110, 112, 143; intergenerational relationships, 102, 103, 107, 111; municipal paperwork, 112; negative impacts of, 60–2; partnerships, 107, 111–12; place and space, 108, 111–12; practitioners' roles, 111–12; research needed on, 32, 101, 112; rewards and benefits, 39–40, 107–8; safety concerns, 108–9; sizes of, 102–3, 105, 111, 126; social capital, 69, 87, 110; time span, 32–3; types of, 102–6, 108; unconventional services, 151–2. *See also* art fairs and festivals; health fairs and festivals

farmers' markets, 88, 109

fashion festivals, 46, 117

female. *See* gender

Festa del grillo, Italy, 46

Festival Internacional de Cine de Valdivia, Spain, 20

festivals: about, 31, 114–15, 123, 129, 154–6; attendees, 35, 121, 124–5, 126; branding of communities, 31, 116, 119; capacity enhancement, 120; case study (Lunar New Year Festival, SF), 155–63; classification of, 38–9, 47, 126; combined events, 71–2, 87, 117, 119, 122, 158–9, 160; cultural capital, 31, 116, 118–19, 124; definitions of, 47, 115; economic capital, 30–1, 119, 122, 123, 127; environmental issues, 126–7; evaluation of, 79, 81–2, 123–5; failed events, 122; and food, 15–16, 46, 116, 117, 126; goals and objectives, 116, 119, 120, 122, 126, 127; green spaces, 88; historical background, 114–15, 127; intergenerational events, 47; literature on, 30–1, 79, 114–17, 125–6; media coverage, 154; negative impacts of, 60–2; partnerships, 120; place and space, 86–7, 115, 117, 118–19; political folklore, 119; practitioners' roles, 125–9; recent trends, 31; research needed on, 32, 116, 125; rewards and benefits, 39–40, 118–21, 124, 154; seasonal preferences, 117; sizes of, 120, 122, 126; and social capital, 117, 120; and social justice, 47, 116; staff, 122; stakeholders, 120, 122, 127; statistics on, 117; and sustainability, 121–2; themes of, 47; time spans, 25, 32–3, 117–18; and tourism, 31, 48, 114, 116, 157–63; types of, 46, 115–18, 126; volunteers, 120–1. *See also* art fairs and festivals; health fairs and festivals; music

field-configuring events, 26. *See also* celebrations

La Fiesta, Washington DC, 118

Fiesta Boricua, Chicago, 132

fiestas, as term, 114. *See also* festivals

film festivals, 116, 117, 125–6
Fine, G.A., 6
fires and fireworks, 57, 58
flower festivals, 117, 158, 162
Foner, N., 145
food: about, 25; alcohol, 54–6, 122; community gardens, 10–11, 13, 36, 37, 87–8; cultural competence and humility, 94–5; as economic capital, 20; at fairs, 101, 104, 106, 109; at festivals, 15–16, 46, 116–17, 124, 126; food-related festivals, 15–16, 46, 117; food safety, 56, 109; gardening festivals, 46; at school events, 124; traditional foods, 20; wine festivals, 117
France, music festivals, 52
fun and play: about, 36; and food, 25; and fundraising, 39; as human right, 90; motivation of attendees, 33, 36; and photography, 78, 81
funding and sponsorships: about, 15, 75–6, 189; accountability in, 15, 75–6; evaluation of economic capital, 72–4; evaluation to access, 63; failed events, 122; of fairs, 107; goals and objectives, 159–60; of parades, 132, 134, 136; and political tensions, 136; practitioners' roles, 76; recent trends, 15; sponsoring organizations, 193; sponsorships by corporations, 15, 55–6, 75–6, 189; sponsorships by local businesses, 189; tobacco industry sponsors, 55. *See also* economic capital
funerals. *See* memorials

Gabon, Independence Jubilee, 122
gardening festivals, 46

gardens, community, 10–11, 13, 36, 37, 87–8
gay community. *See* LGBTQ communities
gender: attendees, 35, 68; beauty pageants, 159–62, 170; evaluation of events, 68; male privilege and sexism, 35; motivation of attendees, 35, 73–4; and power relations, 35–6; public space as heterosexual, 172; queer women's activism, 172; race and ethnicity, 68; of stakeholders and leaders, 35, 68. *See also* LGBTQ communities
gentrification of urban areas, 52, 166, 167–8
genuineness. *See* authenticity
geographical place. *See* place and space
Germany, festivals and parades, 15–16, 50–1, 135
Getz, D., 28, 38–9, 114–15, 120
Gieseking, J.J., 9
Girth and Mirth, 12
global migration. *See* newcomers
goals, objectives, and purposes: about, 14, 39, 192; branding of neighborhoods, 14–15, 70; community development, 14–15; community participation, 90–2; Delphi technique, 81–2; economic development, 13; economic gain, 16, 30–1, 70; education of outsiders, 58; and evaluation, 70, 74, 124; failed events, 122; of fairs, 101–2, 105, 112–13; of festivals, 119, 124, 126, 127–8; literature on, 33; of marginalized communities, 145; motivation of organizers, 33–6; multiple local needs, 30, 36,

68; and political tensions, 127–8; purposes of celebrations, 29–33; rewards and benefits, 39–40; social capital increase, 11, 87; social change, 31, 89–90; social identity, 14; stages in, 97–8; and values, 84–5, 88–9. *See also* empowerment and ownership

Goldblatt, J.J., 27–8, 72, 89

Grand Mass Gymnastics, North Korea, 122

gravitational force of events, 34–5, 81

The Great Good Place (Oldenburg), 15–16

green movement. *See* environmental issues

green spaces, 27, 87–8. *See also* gardens, community

groups. *See* age groups; attendees; crowds

Haitian communities, 71–2

Harvey, D., 89

Hayden, D., 86

health fairs and festivals: about, 103, 105, 110–11, 143–4, 153; attendees, 105, 108, 110–11, 150–1; case study (Chinatown's Health Fair, NYC), 143–53; community engagement, 146–7; definition of, 149; evaluation of, 110–11; follow-up visits, 149; goals and objectives, 149–50; language and literacy services, 105, 146, 147, 151–2; media coverage, 152–3; medical discoveries, 108; medical staff for, 152; mental health events, 116, 148; for newcomers, 105, 151; service-learning projects, 106–7; and social justice, 146, 150–1; unconventional services, 151–2; undocumented newcomers, 151

health issues: about, 54–9, 72–3; aggressive behaviors, 54–6; alcohol and illicit drug use, 54–6, 122; animals at events, 109; communicable diseases, 58; community media coverage, 152–3; crowd control, 54–5, 57, 135; environmental hazards, 57, 58; evaluation of, 72–3; event medicine, 57–8, 135; fires and smoke, 56, 57, 58; food-borne illnesses, 56, 109; noise, 25, 57, 168; at parades, 135–6; risky sexual behavior, 56, 58; stress-related illnesses, 59; suicides, 59. *See also* safety issues

heritage assets. *See* cultural capital

higher education. *See* academic institutions; academic research

higher education students. *See* youth and young adults

Hindu festivals, 56, 57

Hispanic/Latino communities: Cinco de Mayo, 55–6; fiestas (festivals), 114; marginalized communities, 170–1; memorial events, 46; music fairs for youth, 104; parades, 132, 164, 169–71, 173

history of a community, 49, 52–3, 80, 119, 145, 180–1

homosexual. *See* LGBTQ communities

human capital: about, 19–20, 194; capacity enhancement, 87–8; changes in, 23; children and youth, 22, 92; education and learning, 22–3; fairs, 106; festivals, 120; and green spaces, 87–8; and health,

150, 152; and inclusiveness, 59; intergenerational learning, 20, 22–3; literature on, 19; volunteers, 48. *See also* capital/assets framework; children; health fairs and festivals; leadership; volunteers and civic engagement; youth and young adults

human services: case study (Chinatown's Health Fair, NYC), 143–53; evaluation of events, 110; fairs, 53, 101, 103, 104, 105, 110, 112; social networks and providers of, 53. *See also* health fairs and festivals; stakeholders

Humm-Delgado, D., 96–7

Hurricane Katrina, 133

identity and celebrations: about, 44, 73; and authenticity, 42–4; and changing demographics, 52; and community development, 44; cultural competence and humility, 94–5; evaluation of, 72–3; LGBTQ identity, 172–4; literature on, 85; national events, 164; newcomers, 85; parades as performance of, 131, 139; and place, 9, 86; and rituals, 41–2; sexual identity, 74, 165; visible minorities, 7. *See also* branding of neighborhoods; LGBTQ communities

Iltis, A., 41–2

immigrants. *See* newcomers

inclusion/exclusion politics, 76–7.

See also excluded communities

Independence Jubilee, Gabon, 122

India, religious festivals, 47, 57

Indigenous culture and people, 173

insect festivals, 46

intangible capital: about, 19, 22, 69; community gardens, 88; cultural value of artifacts, 22; evaluation of, 69, 123–4. *See also* capital/assets framework

intergenerational relationships. *See* age groups

Irish communities, 133

Israeli apartheid, 173

Italian communities, 146, 156, 165, 167

Italy, festivals, 46

Jackson, M.J., 105

Jaimangal-Jones, D., 191

Jamaica, funerals, 46

James, J., 72, 96

James, P., 66, 91

Jinan, China, 57

job fairs, 104

Kazin, A., 7–8

Kretzmann, J.P., 13, 85

Ku Klux Klan, 59

Labor Day Parade, Michigan, 86

Lagos, Nigeria, 7, 56

Landers, F., 36

languages: about, 185; and acculturation continuum, 147, 151, 181–2, 185; community characteristics, 147; at health fairs, 105; labeling and stereotypes, 93; and media, 182; practitioners' skills, 185

Latin America: Cinco de Mayo, 55–6; Día de los Muertos, 46; plazas as public places, 21. *See also* Hispanic/Latino communities

leadership: about, 95–6, 120; academic support for, 96;

and authenticity, 43; capacity enhancement, 85, 87–8; development as benefit, 39; gender issues, 35, 68; as investment, 95–6; practitioner's role, 12; sponsoring organizations, 193; stakeholders, 120, 122. *See also* human capital; stakeholders

learning. *See* education and learning

leisure. *See* fun and play

Lekies, K.S., 29

lesbian community. *See* LGBTQ communities

LGBTQ communities: changes to stereotypes, 11–12; children at events, 128–9; combined events, 164, 172; demographics of attendees, 35–6; diversity within, 128, 172–4; evaluation of attendance, 73–4; evaluation of events, 191; events as rituals, 41–2; health and safety issues, 58; intersectionality of issues, 128, 133, 172–4; motivation of attendees, 35–6, 73–4; political empowerment, 35–6, 41, 128; and political tensions, 136; politicization of identity, 172–4; pride events, 35–6, 128–9, 164–5, 172–5; pride parades, 23, 41, 59, 128–9, 164–5, 172–5; public space and sexual identity, 172; queer women's activism, 172; segment exclusion, 12, 76–7, 133; and social justice, 165, 173–4. *See also* excluded communities

libraries, 45–6, 183. *See also* records and archives

liminality of celebrations, 29–30

literacy programs, 151–2, 155

literary fairs and festivals, 103, 117, 151–2

local control. *See* empowerment and ownership

Los Angeles, Chinatown, 144–5

Loukaitou-Sideris, A., 6, 27, 130

Love Parade, Germany, 135

loyalty of attendees, 35, 52. *See also* attendees

Lunar New Year celebrations, 6, 57, 155, 158. *See also* Chinese community, San Francisco

Macy's Thanksgiving Parade, 137

Mair, J., 9, 44, 64

male. *See* gender

Mangold, W., 9

Mardi Gras Indian Tribes, New Orleans, 133

marginalized communities: about, 188–9; capacity enhancement, 89–90, 97–8; cultural competence and humility, 94–5; deficit paradigm, 7–9; diversity within, 38, 94–5; empowerment of, 92–3; evaluation of events, 68, 71, 72–4; goals of events, 68, 120, 145; historical projects, 49, 52–3; history of the community, 180–1; multiple stigmas, 50; records and archives, 49; research needed on events, 32; rewards and benefits of events, 39–40; segregated communities, 7; service-learning projects, 121; volunteers at events, 121. *See also* communities; excluded communities; LGBTQ communities; race and ethnicity; social justice; Toronto, Ontario, parades

marijuana fairs, 110
marketing and sales: and authenticity, 43; evaluation of, 67–8; failed events, 122; marketing of tobacco products, 55; and participant motivation, 125; and social media, 81, 183; and technologies, 108, 161–2
Maslow's hierarchy of needs, 48
Massachusetts: joint events and communities, 71–2; St Patrick's Day parades, 133, 136, 139–40
McKay, S.L., 35
McKnight, J., 13, 35, 78, 85
media coverage: about, 32, 152–3; crowd estimates, 54; deficit bias, 8–9; of festivals, 154; and gender, 161; of health issues, 152–3; by local ethnic media, 32, 152–3; of parades, 132–3, 134–5, 137, 139; and race and ethnicity, 161; as record of event, 184; social media, 108, 183, 184–5. *See also* branding of neighborhoods
media relations: about, 182, 184–5; capacity enhancement, 87, 184–5; local ethnic media, 32, 152–3; media literacy and group learning, 45; practitioners' knowledge and skills, 182, 184–5; and social media, 108, 183, 184–5
mega-events, 28–9, 76, 103, 122
memorials, 16, 46–7, 52. *See also* rituals
memory: collective memories, 6, 9, 16, 21, 37, 49, 51, 111; cultural authenticity, 42–4, 49; environmental autobiographies, 9; as last stage, 45; librarians' roles, 45–6; memorials, 16, 46–7, 52; place and identity, 9; records and archives, 59–60; as social anchor, 37; and transnational communities, 48–9. *See also* authenticity; newcomers; records and archives
men. *See* gender
"meso-sociological" approach, 6
Mexican-American communities: children's science fairs, 104; Cinco de Mayo, 55–6, 115; parades, 132; youth carnival drum fairs, 104
Michigan, 58, 86
micro-enterprises. *See* small businesses
military parades, 131
Miranda, M., 36
Moeran, B., 32
Montreal, Quebec: marginalized communities, 94; writing communities, 44–5
Moss, S., 69
motivation of attendees: about, 5–6, 33–6, 73–4; categorization of, 35; emotional connection, 34; evaluation of, 73–4, 125; expectation of authenticity, 42; fun and play, 33, 36, 73–4; gravitational force of events, 34–5, 81; homecomings and reunions, 34; learning, 45; literature on, 35; racial and ethnic groups, 34; research needed on, 33–4; sociodemographic factors, 35–6. *See also* attendees; fun and play
motivation of organizers: about, 33–6, 193; expectation of authenticity, 42; literature on, 33; Maslow's hierarchy of needs, 48; political goals of stakeholders, 33; research needed on, 48; rewards and benefits, 39–40. *See also* goals, objectives, and purposes

multigenerational relationships. *See* age groups

museums, 45, 159

music: capacity enhancement, 3–4; classification of events, 29, 47; combined events, 117, 173–4; as community asset, 18–19, 30, 104; crowd surveillance, 108; cultural competence and humility, 94–5; fairs, 103, 104; festivals, 30, 52, 117, 118, 126; gender of attendees, 35; LGBTQ pride events, 173–4; literature on, 117; motivation of attendees, 35, 52; parades, 173–4

national days, 164

national identity. *See* identity and celebrations

National Mall, Washington DC, 118

New Orleans, parades and festivals, 127, 133, 138

New Year's celebrations, 6, 57, 158. *See also* Chinese community, San Francisco

New York City: Asian newcomers, 144–6; bicycle protest parades, 131; case study (Chinatown's Health Fair), 143–53; Charles B. Wang Community Health Center, 143–4, 146–53; children, 51; Chinatown's history, 144–9; ethnic parades, 131; health fairs, 149; Little Italy, 146; Macy's parade, 137–8; Mexican-American community, 51; newcomers, 51, 145; Persian parades, 131, 138; Pride Parade, 128; Puerto Rican community, 5, 36; St Patrick's Day parades, 133, 136, 139–40. *See also* Chinese community, New York City

newcomers: about, 48–53; changes to stereotypes, 11–12; changing demographics, 52; children and youth, 50–1, 92; fairs for, 105, 108; festivals as bridge between old and new, 51, 114–15; goals of events, 145, 155; and green spaces, 87–8; historical projects, 52–3; history of the community, 180–1; homecoming and reunions, 34; inclusion of, 50–3, 91; multiple stigmas, 50; national events, 164; national identity, 85, 164; organizations for support, 50–1; parades, 168; place and belonging, 145, 171; political tensions, 49–50, 145; resilience of, 18, 49–50; and social capital, 11, 20; undocumented newcomers, 32, 60, 74, 109, 151. *See also* excluded communities; languages; marginalized communities

news. *See* media coverage; media relations

Nigeria: deficit paradigm, 7; festivals, 40–1, 56

No Fool Is Illegal festival, Germany, 50–1

noise issues, 25, 57, 168

North Korea, gymnastics, 122

Northern Ireland, 35, 41, 57–8

objectives for celebrations. *See* goals, objectives, and purposes

Ohio, community gardens, 10–11

Oje'nmeho Cultural Festival, Nigeria, 40–1

Oldenburg, R., 15–16

older adults: ageism, 59; local ethnic media, 152–3; safety concerns, 55. *See also* age groups

Ontario. *See* Ottawa, Ontario; Toronto, Ontario
oral tradition. *See* storytelling
Orange Parade, Northern Ireland, 41
organic, 42. *See also* empowerment and ownership
original culture. *See* authenticity
Ottawa, Ontario: Chinatown's New Year Festival, 6
ownership, community. *See* empowerment and ownership

parades: about, 130–1, 140, 164, 175; anti-social behavior, 59, 135–6; artistic talent, 86, 134; case studies on, 79, 175; classification of, 38–9, 47; combined events, 71–2, 87, 119, 158–9, 160, 172; crowds at, 25, 30; cultural identity, 86, 139; definition of, 130–1; disruptions from, 30, 132–3; evaluation of, 75, 79, 137; floats, 134, 139; goals and objectives, 130, 164; identity performance, 131; marketing of, 139; media coverage of, 132–3, 134–5, 137, 139; negative impacts of, 60–2; newcomers, 168; noise issues, 25, 168; place and space, 86–7, 130, 132, 139, 175; political capital, 133–4, 170–3; political tensions, 127–8, 134–7, 140; practitioners' roles, 138–40; pride parades, 23, 41, 59, 128–9, 164–5, 172–5; race and ethnicity, 131–3; rewards and benefits, 39–40, 132–4; and rituals, 41–2; routes of, 30; and social justice, 128, 131, 173–4; sponsorships, 134; as territorial acts, 27; time span, 26, 32–3, 139; types of, 131. *See also* Toronto, Ontario, parades

participant observation, 79, 137
participants in celebrations. *See* attendees
partnerships: about, 31–2, 47, 97; as community asset, 13; economic capital, 31–2; education and learning, 111–12; evaluation of, 123, 124; and fairs, 107, 111–12; of neighborhood associations, 11; practitioners' roles, 111–12; public-private partnerships, 123; stakeholders, 120, 122
Pasadena, California, 137
Pedersen, J.S., 32
performing arts festivals, 22
Persian parades, New York City, 131, 138
photography and evaluation, 77–8, 81, 137
physical capital: about, 19–21; capacity enhancement, 87–8; collective memories, 21; community gardens, 87–8; literature on, 20–1. *See also* capital/assets framework; place and space
place and space: about, 11, 16–17, 20–1, 86; ambiance, 37–8, 77; arts and development of, 16; assets perspective, 16; and authenticity, 42–4; capacity enhancement, 86–8; in capital/assets framework, 19–21; case studies on, 79; ethnic-specific places, 38; evaluation of, 79; and fairs, 118; and festivals, 114, 117, 118, 155; gentrification, 167–8; green spaces, 27, 87–8; as heterosexual, 172; and identity, 9; literature on, 21; memorials, 16, 46–7, 52; newcomers, 38, 51, 171; parades in, 27, 86, 87, 169;

292 Index

privatization of public space, 14, 16, 60; public spaces, 6, 16, 41–2, 118, 155; recent trends, 14; research needed on, 21; ritual spaces, 41–2; segment exclusion, 76–7; site selection, 79; social anchors, 37; and social capital, 11, 21, 87; streets as, 87, 104, 118, 158–9; urban enclaves, 52; urban ethnology, 10. *See also* gardens, community; physical capital

Planning Community-Wide Special Events, 7

play and fun. *See* fun and play

political capital: about, 19–20, 72; community gardens, 87–8; evaluation of, 72–3; and festivals, 119, 127–8; intersectionality of issues, 172–3; local negotiations, 20; and parades, 133–4, 170–3; political goals, 128; political goals of stakeholders, 33; public policy agendas, 44. *See also* capital/assets framework; volunteers and civic engagement

political tensions: and economic motives, 55, 127; in host communities, 55; at LGBTQ events, 128–9; and newcomers, 49–50, 145; in Northern Ireland, 41, 57–8; and parades, 59, 127–9, 134–7, 140; privatization of public space, 14, 60; public policy agendas, 44; stakeholders with agendas, 33, 127; in transnational communities, 49–50, 53, 159; as urban cracks, 37. *See also* empowerment and ownership

Portland, Oregon: Chinatown's history, 144–5

Portugal, 166, 168–9

Portuguese Day Parade, Toronto, 166–9

post-secondary education. *See* academic institutions

power relations. *See* empowerment and ownership

practitioners' knowledge and skills: about, 179–80, 183–7; case studies for, 79; cultural competence and humility, 94–5; goals of this book, 24, 179–80; group and interpersonal skills, 185–6; history of the community, 180–1; languages, 185; media relations, 32, 152–3, 182, 184–5; principles for, 96–7; professional community, 186; records and archives, 183; specializations, 179, 192–3; urban ethnology, 10; values and traditions, 180–2; volunteers, 182–3. *See also* case studies

practitioners' roles: about, 12, 96–7; access to resources, 39, 76, 96–7; cultural and social brokers, 12, 112, 138; evaluation of process and outcomes, 67, 138; at fairs, 111–12; at festivals, 125–9; funding and sponsorships, 138; goals of this book, 24; labeling and stereotypes, 93; leadership development by, 12, 95–6; negotiations, permits, and paperwork, 112, 138; at parades, 138–40; partnerships, 111–12; principles for, 96–7

pride in communities. *See* cultural capital

pride parades. *See* LGBTQ communities

principles and values. *See* values and traditions
principles of democracy. *See* democratic principles
print media. *See* media coverage; media relations
prison fairs, 104
public relations. *See* media relations
public space. *See* place and space
Puerto Rican community, 5, 10, 36, 132
purity. *See* authenticity
purposes for celebrations. *See* goals, objectives, and purposes
pyramid contests, 56

Quebec, writing communities, 44–5
queer community. *See* LGBTQ communities
Queers Against Israeli Apartheid, 173

race and ethnicity: animals at events, 109; Australian Indigenous festivals, 117; branding to lessen tensions, 160; changes to stereotypes, 11–12, 90, 160, 162; combined events, 71–2; community tensions, 128, 160; ethnic-specific places, 38; evaluation of events, 68, 72–3; excluded communities, 120; gender issues, 68; history of the community, 180–1; identity and celebrations, 7; intolerance in parades, 59; local ethnic media, 152–3; motivation of attendees, 34, 73–4; nationality festivals, 128; parades, 59, 131–3; segment exclusion, 76–7; and sustainability, 90; traditional foods, 20; and urban cracks, 37. *See also* excluded communities; newcomers; political tensions; stereotypes; *and entries beginning with* Chinese communities
radio. *See* media coverage; media relations
Ray, B., 94
real events. *See* authenticity
records and archives: about, 45–6, 59–60, 183; access to, 78; digital technologies, 77–8; evaluation reports, 82–3; event institutionalization, 59–60; historical projects, 53; librarians' roles, 45–6, 183; of marginalized community histories, 49; media coverage, 184; photography, 77–8, 81, 137; practitioners' knowledge and skills, 183; social media, 184–5; storage of, 78, 183; types of, 78
recreation. *See* fun and play
refugees. *See* newcomers
religion and religious events: attendees, 23, 35; combined events, 117; fairs and festivals, 103, 117, 118; Hindu festivals, 56, 57; impact on stereotypes, 58; intolerance in parades, 59; literature on, 117; media sources, 182; partnerships for events, 13; place and space, 118; political tensions, 41, 57–8; role in events, 25; stampedes, 57
renaissance fairs, 103
reports. *See* records and archives
resilience, 18, 49–50
Rio de Janeiro, carnivals, 108, 122
risks. *See* safety issues
rituals: about, 40–2; literature on, 40–2; memorials, 16, 46–7; ritual spaces, 41–2

Rose Day Parade, Pasadena, 137
Rosendahl, T.J., 128, 173–4
Rosenstein, C., 21–2
rural areas, 8, 38. *See also* county and country fairs

Sacramento, California, Chinese community, 144–5
safety issues: age groups, 55; anti-social behavior, 127, 135–6; children, 55, 57, 58, 109; crime, 154; crowd control, 54–5, 57, 135; at fairs, 108–9; at festivals, 129; food-borne illnesses, 56, 109; in LGBTQ community, 58, 129; noise issues, 25, 57, 168; older adults, 55; at parades, 135–6; risky sexual behavior, 56. *See also* alcohol and illicit drugs
St Patrick's Day parades, 133, 136, 139–40
sales. *See* economic capital; marketing and sales
same-sex relationships. *See* LGBTQ communities
San Francisco, California: Italian community, 156; tourism, 157–63. *See also* Chinese community, San Francisco
Santiago, Chile, LGBTQ events, 129
sausage festivals, 15–16
Scerri, A., 66, 91
Schippers, H., 29
schools: fairs, 103, 104, 105, 111–12; festivals, 120; food at events, 124; historical projects, 53; instruction on local culture, 53; partnerships, 111–12, 120, 124; place and space, 118; service-learning projects, 106–7, 111, 121; student participation, 68; teacher participation, 53. *See also* education and learning
science fairs, 103, 104
Scotland, Edinburgh Festival, 22
Seattle, Washington, Chinese community, 144–5
segment exclusion, 76–7. *See also* excluded communities
segregated communities, 7
Serbia, LGBTQ events, 129
service-learning projects, 106–7, 111, 121
sexual behavior, 56, 58, 59
size of event: about, 28–9, 192; estimates of crowds, 54, 108; of fairs, 102–3, 105, 111; of festivals, 126; and goals, 192; large-scale exhibitions, 29; mega-events, 28–9, 76, 122; small events, 11, 26, 87, 126, 192; and social capital, 11
small businesses: as community asset, 13, 189; fairs, 103, 105–6; festivals, 156; participation as ritual, 40–2; partnerships as asset, 13; street fairs, 104, 158–9. *See also* economic capital; partnerships
small-scale events, 11, 26, 87, 126, 192. *See also* celebrations; size of event
Smith, A., 76
Smith, J.P., 28, 71–2
smoke and smoking, 55, 56
Social Aid and Pleasure Clubs, New Orleans, 133
social anchors, 37
social capital: about, 11, 19–20, 72–4; changes to stereotypes, 11–12; community gardens, 10–11, 87–8; evaluation of, 68–9, 72–4, 110; and fairs, 69, 110; and festivals, 117, 120; fun and play, 36; function of

celebrations, 6, 8; literature on, 19, 20; memorial events and bonding, 47; and newcomers, 11, 20; and resilience, 18, 49–50; rituals and social connections, 41–2; and social anchors, 37. *See also* capital/assets framework

social change. *See* social justice

social justice: about, 87, 89–90; activism against Israeli apartheid, 173; arts activism, 87, 92; capacity enhancement, 87, 89–90; collective memories, 21; and empowerment, 92–3; and evaluation, 69; and festivals, 47, 116; film festivals, 116, 117, 125–6; fun and play, 36; and health fairs, 149–51; homelessness, 117; and marginalized communities, 69, 89; and parades, 128, 131, 173; partnerships with organizations, 111–12; and place and space, 21; service-learning projects, 111; and sustainability, 90; urban citizenship, 91–2; urban cracks, 37; water and sanitation, 46, 47. *See also* empowerment and ownership; excluded communities; LGBTQ communities; marginalized communities

Social Justice and the City (Harvey), 89

social media, 81, 108, 111, 183, 184–5. *See also* technology

Social Work Practice in Nontraditional Urban Settings (Delgado), 13

social workers. *See* practitioners' knowledge and skills; practitioners' roles

space. *See* place and space

Spain, festival, 20

special events, 26. *See also* celebrations

sponsorships. *See* funding and sponsorships

sports events: as bridge between old and new cultures, 114; capacity enhancement, 87; as celebrations, 44; Chinese New Year Run, 159; place and space, 118; as type of celebration, 38

stakeholders: about, 120; determination of authenticity, 43; gender issues, 35, 68; leadership, 95–6, 120, 122; motivation of, 33–6; political tensions, 33, 127. *See also* human capital; leadership

stampedes, 57

stereotypes: ageism, 59; beauty pageants, 160–1; and labeling, 93; mental health arts festivals, 116; and pride parades, 59; race and ethnicity, 90, 160, 162; and religious celebrations, 58; research to avoid, 107

storytelling: age groups, 181; case studies as, 79; and digital technologies, 77–8; as evaluation, 66, 69–70, 77–8; historical projects, 53, 180–1; and photography, 77–8, 81; place and space for, 21; urban legends, 58

street parties and fairs, 87, 104, 118, 158–9, 162. *See also* fairs

students. *See* children; schools; youth and young adults

suicides, 59

sustainability: about, 90–1; and authenticity, 43; capacity enhancement, 90; combined

events, 71–2; creative approaches, 71–2; and evaluation, 91; and excluded communities, 129; and fairs, 107, 110; and festivals, 121, 126–7; health fairs, 148; literature on, 91; and social justice, 90
Sydney, Australia, LGBTQ events, 58

talent contests and fairs, 45, 56, 103
Tam, Thomas, 148
teams and celebrations. *See* partnerships
technology: about, 80–1; crowd sensing and simulations, 54, 108; cyber-communities, 10; digital screens and crowds, 54; and evaluation, 77–8, 80–1; mobile devices, 54; mobile positioning (GPS), 80–1; and parades, 30, 139; records and archives, 78; social media, 81, 108, 111, 183, 184–5; video ethnography, 77–8
television. *See* media coverage; media relations
theater festivals, 22, 117
Thibaud, J.P., 37–8
Tibet, tourism, 85
Time Maps (Zerubavel), 49
tobacco, 55, 56
Toilet Festival, India, 47
tolerance. *See* stereotypes
Toronto, Ontario: beauty pageants, 170; case studies, 165, 175; Chinatown's history, 145, 166; demographics, 165, 167–8, 169–70; economic capital, 155, 172, 174; festivals, 155, 170; gentrification, 167–8; Hispanic/Latino community, 164, 169–71, 173; historical background, 165–7, 169–70; Italians, 165, 167; LGBTQ community, 165; Multicultural Festival, 170; newcomers, 165–7, 170–1; place and space, 167, 169; Portuguese community, 166–9; Pride Week, 164–5, 172–5; small businesses, 166, 170, 174
Toronto, Ontario, parades: about, 164–6, 175; activism against Israeli apartheid, 173; Anthony Day, Little Italy, 165; Chinese New Year, 165; Hispanic Day Parade, 170–1; Hispanic/Latino, 164, 169–71, 173; intersectionality of communities, 173; Italian Parade, 165; LGBTQ pride parades, 164–5, 172–5; Portuguese Day, 164, 166–9; Trinidad and Tobago Festival, 165
tourism: about, 48; and authenticity, 42–4; and branding of communities, 14; capacity enhancement, 85; Chinese celebrations, 157–63; commercialism, 15–16, 43–4; economic capital, 18, 43–4, 48, 127; evaluation of economic benefits, 68–9, 70; fairs to attract, 105; and festivals, 31, 114, 125–6, 127; film festivals, 125–6; as goal, 159–60; literature on, 48; motivation of tourists, 34; recent trends, 16; urban tourism, 47–8
trade fairs, 103
traditional arts festivals, 117
transgender community. *See* LGBTQ communities
transnationalism. *See* newcomers
treasure hunts, 159
Trinidad and Tobago Festival, Toronto, 165

true events. *See* authenticity
Turkish festivals, Chicago, 128

UN Convention on the Rights of the Child, 90
undocumented newcomers, 32, 60, 74, 109, 151
United Kingdom, volunteers, 120–1
universities and colleges. *See* academic institutions; schools
university students. *See* youth and young adults
urban communities. *See* communities; place and space
urban cracks, 37
urban enclaves, 52
urban ethnology, 10
urban legends, 58. *See also* storytelling
urban tourism, 47–8

values and traditions: about, 84–5, 88–9, 98, 181–2; and capacity enhancement, 84–5; community assets, 18–19, 93–4; community investment, 95–6; community participation, 90–2; empowerment, 92–3; and evaluation, 63, 65–6, 72, 123; and festivals, 114; history of the community, 180–1; newcomers, 181; practitioners' knowledge of, 94–5, 180–2; practitioners' principles, 96–8; research needed on, 89; social justice, 89–90; use of local knowledge, 95; value clarity, 89. *See also* authenticity; newcomers; social justice
van Heerden, E., 29–30
Vancouver, Canada, Chinese community, 145

vendor fairs, 103
venues. *See* place and space
Veronis, L., 171
videography and evaluation, 77–8, 81
violence. *See* anti-social behavior
Vishu festival, India, 57
visible minorities. *See* excluded communities; marginalized communities; race and ethnicity
visitors at celebrations. *See* attendees
visual ethnography, 77–8
Viva Chicago parade, 132
volunteers and civic engagement: about, 48, 70–1, 182–3; and evaluation, 70–1, 74, 78; and fairs, 107, 109; and festivals, 120–1; inclusive strategies, 183; Maslow's hierarchy of needs, 48; motivation of, 33–6, 86, 121; photography, 78; as political capital, 20, 86; practitioners' knowledge of, 182–3; research needed on, 48; segment exclusion, 76–7; work hours, 109; youth and young adults, 10–11, 22, 120–1. *See also* human capital

Walmsley, B., 69
Wang Community Health Center, New York City, 143–4, 146–53
Ward, Colin, 17
Washington, Chinese community, 144–5
Washington DC, festivals, 118
water activities, 118
wellness, community, 18, 90. *See also* capital/assets framework; health fairs and festivals; health issues; social justice
Whitford, M., 9, 30, 44, 64
wildlife festivals, 46

wine festivals, 117
women. *See* gender
Wong, B.P., 156
Word on the Street, Toronto, 155
writing activities, 44–5

youth and young adults: classification of events, 29; community gardens, 10; at fairs, 102–3, 104, 107–8, 111; as human capital, 22, 92; intergenerational relationships, 22–3; languages, 181–2; motivation of attendees, 35; newcomers, 50–1, 181–2; participation of, 90, 92; and photography, 78; records and archives, 183; service-learning projects, 106–7, 111, 121; talent contests, 45, 56; as volunteers, 10–11, 22, 120–1; youth culture, 181. *See also* age groups; schools

Zerubavel, E., 49